Philosophy and Medicine

Volume 118

The Philosophy and Medicine series is dedicated to publishing monographs and collections of essays that contribute importantly to scholarship in bioethics and the philosophy of medicine. The series addresses the full scope of issues in bioethics, from euthanasia to justice and solidarity in health care. The Philosophy and Medicine series places the scholarship of bioethics within studies of basic problems in the epistemology and metaphysics of medicine. The latter publications explore such issues as models of explanation in medicine, concepts of health and disease, clinical judgment, the meaning of human dignity, the definition of death, and the significance of beneficence, virtue, and consensus in health care. The series seeks to publish the best of philosophical work directed to health care and the biomedical sciences.

More information about this series at http://www.springer.com/series/6414

Gérard Reach

The Mental Mechanisms of Patient Adherence to Long-Term Therapies

Mind and Care

 Springer

Gérard Reach
Avicenne Hospital
    and Paris 13 University
Sorbonne Paris Cité
Bobigny
France

Translation by Nastya Solovieva, edited with the assistance of Jeff Engelhardt and John Meyers

ISSN 0376-7418   ISSN 2215-0080   (electronic)
Philosophy and Medicine
ISBN 978-3-319-38541-9   ISBN 978-3-319-12265-6   (eBook)
DOI 10.1007/978-3-319-12265-6

Translation from the French language edition: Pourquoi se Soigne-t-on, Enquête sur la Rationalité Morale de l'Observance, © Le Bord de l'Eau 2007. All rights reserved.

Springer Cham Heidelberg New York Dordrecht London
© Springer International Publishing Switzerland 2015
Softcover reprint of the hardcover 1st edition 2015

Printed on acid-free paper

Springer International Publishing AG Switzerland is part of Springer Science+Business Media (www.springer.com)

'O Socrates, Gorgias is deceiving you, for my art is concerned with the greatest good of men, and not his.' And when I ask, Who are you? He will reply, 'I am a physician'. What do you mean, I shall say. Do you mean that your art produces the greatest good? 'Certainly' he will answer, 'for is not health the greatest good? What greater good can men have, Socrates?'

Plato, *Gorgias*

For neither does the man who is ill become well on those terms, although he may, perhaps, be ill voluntarily, through living incontinently and disobeying his doctors. In that case it was then open to him not to be ill, but not now, when he has thrown away his chance, just as when you have let a stone go it is too late to recover it; but yet it was in your power to throw it, since the moving principle was in you.

Aristotle, *Nicomachean Ethics*

You will see, I will try to show you, how generally speaking the principle that one must take care of oneself became the principle of all rational conduct in all forms of active life that would truly conform to the principle of moral rationality.

Michel Foucault, *The Hermeneutics of the Subject*

I am a sick man. I am a wicked man. An unattractive man. I think my liver hurts. However, I don't know a fig about my sickness, and am not sure what it is that hurts me. I am not being treated, and never have been, though I respect medicine and doctors. What's more, I am also superstitious in the extreme; well, at least enough to respect medicine...No, sir, I refuse to be treated out of wickedness. Now, you will certainly not be so good as to understand this.

Dostoyevsky, *Notes from the Underground*

"He" has two antagonists; the first presses him from behind, from his origin. The second blocks the road in front of him. He gives battle to both. Actually, the first supports him in his fight with the second, for he wants to push him forward, and in the same way the second supports him in his fight with the first, since he drives him back. But it is only theoretically so. For it is not only the two antagonists who are there, but he himself as well, and who really knows his intentions? His dream, though, is that some time in an unguarded moment—and this, it must be admitted, would require a night darker than any night has ever been yet—he will jump out of the fighting line and be promoted, on account of his experience in fighting, to the position of umpire over his antagonists in their fight with each other.

Franz Kafka, *Aphorisms*

The force that presses "Him" from behind and the force that blocks Him from the front are the forces of the past and future.

Hannah Arendt, *The Life of the Mind: Thinking*, 1971

To Isabelle
To our children and grandchildren

Foreword

In books on medicine, one generally wonders why one is sick, and how one can look after oneself. One more rarely wonders why one does not take care of oneself, and almost never why one does take care of oneself. The reason is simple: Isn't good health a desirable result in itself? If a patient visits a doctor, isn't it in order for the doctor to take care of him/herself? If one claims to need treatment, isn't it to be provided? Under these conditions, how is it possible for a patient to go to the doctor but nevertheless not take care of him/herself? Yet, there is a plethora of cases in which patients do not follow doctors' prescriptions—patient nonadherence.

This situation is reminiscent of a well-known philosophical paradox, the Socratic paradox:

(1) If somebody wishes X more than Y, and believes that doing A is the best means for him/her to obtain X, and is free to do A, then he/she will do A;

(2) A person wishes X more than Y;

(3) The person does not do A.

Socrates supported the idea that situations such as (3), in which the agent acts against what he/she considers to be the best measure, are impossible because the agent can only be ignorant of what is good, or of what he/she considers best in these cases. "No one is voluntarily malevolent," which is only another manner of stating (1):

(1') Somebody who knows what is good or virtuous to do cannot help but do what he/she considers good or virtuous.

Somebody who goes against (1) is what the Greeks called an *akratès*: Somebody who does not control herself, or what Romans would refer to as incontinent: Those who can say, according to the famous formula *Video meliora, proboque deteriora sequor* (I see what is best, but I do the worst). Socrates (and perhaps Plato) denied the possibility of akrasia. Aristotle, on the other hand, accepted it, claiming that what occurs in the mind of the incontinent individual is undoubtedly a form of bad reasoning: Either he/she does not grasp one of the premises of the practical reasoning well, or does not infer the conclusion correctly. The reasoning in (1)–(3) should obviously be held as follows:

(1*) X is more desirable than Y;

(2*) Doing A will enable me to obtain X;

(3*) I do A.

But assuming that akrasia exists, why doesn't the agent in these cases do A? A frequent answer, which is not incompatible with the Socratic answer, consists in saying that he/she is in the grips of a desire or of such a strong compulsion at the time when he/she would normally be on the verge of doing A. This simply amounts to thinking that he/she "does not control him/herself anymore," and cannot help him/herself. However, it is clear that neither the Socratic answer nor what one can call, according to the American philosopher Donald Davidson, the "principle of Medea" ("I know indeed what evil I intend to do. But stronger than all my afterthoughts is my fury.") gives an explanation for this kind of behavior. If the agent does not know what is good for him/her, he/she is not akratic, but only an ignorant person; and if the agent knows it, but does not do it, one does not understand why he/she adopts this irrational behavior.

The authorities in charge of public health are often faced with this kind of dilemma vis-à-vis certain behaviors such as drinking alcohol before driving or cigarette smoking. They clearly oscillate between pedagogy (Socratic information campaigns to inform bad people where the good is) and the pure and simple constraint (increases in the price of cigarettes, prohibition). Everyone remembers having seen on the TV medical experts despaired to see that drivers do not do what they should judge to be best, namely not to drive after consuming alcohol. If the results of the recent campaigns are very significant, can't there exist a pathway between Socrates and Medea?

Many descriptions of akrasia and weakness of will are provided in philosophy and other literature (in psychoanalytic literature, less so, undoubtedly partly because it uses other names). A vast psychological, medical, economic, sociological, and anthropological literature exists on various irrational behaviors related to akrasia, such as addiction. The phenomenon of patient adherence has been the object of several publications in the field of health psychology. But never, to my knowledge, had this phenomenon been considered originating from the discussions of contemporary philosophers of the mind concerning practical reasoning, the psychology of beliefs and desires, the moral psychology of motivation, and theories of rationality until Professor Gérard Reach's remarkable book, which displays originality in taking these discussions seriously and applying them, with great understanding, to the analysis of the patient-physician relationship. The result is impressive because it represents, to my knowledge, the first true meeting between clinical medicine and the analytical philosophy of the mind and of agency, and for this reason, the model he proposes is of great value for both doctors and philosophers.

A number of works on patient adherence presuppose that the fundamental goal is to achieve adapted behavior on behalf of the patient who does not take care of him/herself, without really considering his/her failure to follow a given step of the treatment to be an action (thus, the product of an intention). Against this implicitly assumed behaviorism, Gérard Reach explicitly defends an approach that can be called "intentionalist," in the sense that it supposes that the patient has mental states referred to as "intentional." This does not simply mean that these mental states, such as beliefs and desires that cause certain behaviors, are the products of an intention, but also that they are endowed with content. On the basis of this intentionalist model, he formulates the problem of nonadherence as a (complex) case of akrasia, and proposes an explanation of this phenomenon that has force as well as subtlety. It gives justice to all the previous literature touching on these subjects, from philosophy to contemporary cognitive science, at the same time having all the empirical force necessary for this kind of investigation. The very nature of his model leads him to treat nonadherence like an action, or rather a series of intentional actions on the part of the agent, that is, the product of a certain choice, and at the same time, like a submissiveness to a set of constraints that the agent does not control, such as habits or emotions. The dilemma of any explanation of irrationality is again an oscillation between the principles of Socrates and Medea. Indeed, one considers akratic behavior—in the medical field, nonadherent behavior— to be the product of either a planned rational action on the part of the agent or of forces beyond the control of the agent. In the first case, the phenomenon disappears: The agent is no longer incontinent since, instead of acting against his/her best judgment, he/she simply changed his/her best judgment and plan. In the second case, the intentionalist model loses its relevance. This situation is what one can fear in the original analysis developed by American psychiatrist George Ainslie, which is based on the phenomenon of time discounting—causing individuals to have a strong preference for the present compared to the future, and to change their preference according to the proximity of the reward.[1]

As shown by philosopher and sociologist Jon Elster, who was early on inspired by the work of Ainslie, the only means available to an agent who—having become aware of the importance for him/her to take care of him/herself—intends to resist a trend that cognitive psychology reveals to be inescapable, is to practice Ulysses' tactics. Ulysses' tactics refers to the use of techniques of precommitment, making reference to Ulysses' request for his sailors to tie him to the mast to avoid yielding to the sirens' songs. In this context, as shown by Gérard Reach, patient education must realize a true inversion of the preferences regarding the present to those regarding the future. To the principle of continence, as per which the agent must achieve the best action according to all the available relevant data, Gérard Reach adds a principle of foresight, according to which the agent (here the *patient*!) should give priority to his/her preferences regarding the future. It is still at the level of intention and of consciousness that this inversion must be carried out, while taking into account which behaviors of the patient can be Medean. Ultimately, time is centric to the patient–doctor relationship. The famous prayer

---

[1] See his book *Breakdown of Will*, Cambridge, Cambridge University Press 2001.

of Saint Augustine—"Give me chastity and continence, but not immediately" (Confessions, VII, 7)—has its therapeutic counterpart: "Give me health and the will to find it or preserve it, but leave me the time to start."

Perhaps one could object that these considerations answer the questions "Why doesn't one take care of oneself?" and "Why does one take care of oneself?" but do they answer the question "Why should one take care of oneself?" or rather "Why should I take care of myself?" that some patients seem to be asking? When one reads astonishing statistics of nonadherence quoted by Gérard Reach (more than 40 % of re-hospitalizations are due to nonadherence) and considers the cases of doctors who drink and smoke, one may wonder whether patients' adherence and what it refers to as the respect of foresight is always, as he suggests, a rational and free choice directed towards the future, with nonadherence being irrational. Certain cases exist wherein it may be that a rational choice is made or caution is exercised by certain patients. One may consider the frequency of nosocomial infections and other serious undesirable events affecting hospitalized patients. In such contexts, the choice of the patient who refuses to take care of him/herself could have certain rational aspects. Conversely, certain patients might make insanity a choice, and the question can arise of whether this choice is thought out and reflected upon. In the movie *Comme une image* by Agnès Jaoui, the character played by Marilou Berry, who obviously has problems with bulimia, is questioned about the reasons for her bad eating habits, and she answers: "It is a whole." *Acedia* (roughly translated as "laziness"), of which Dante speaks in song XVIII of the Purgatory, the inertia of Goncharov's Oblomov, or that of Lord Jim, aren't they also "a whole"? Can't a chronic patient who does not follow his/her treatment make a choice of another kind, like that of a quasi suicide? Can one seek to prevent Serge Gainsbourg from smoking Gitanes or Fidel Castro his Havanes? Isn't the prisoner under the death sentence who, before passing on the electric chair, asks for a light Coke as ironic as Alfred Jarry, who asked for a toothpick on his deathbed? Wouldn't it be necessary for these patients—but also for those who do not take such existential postures—to relearn what Michel Foucault calls "concern for oneself"? This has, in fact, been suggested by some clinicians (particularly in psychiatry).

But the doctor does not have to look after those who have made themselves victims of the tragic sense of life. The same goes for the philosopher, despite everything said by those who want to confine philosophy to training in virtue. The error, as very well said by Gérard Reach, would be precisely to treat the patient like a philosopher, and to provide him/her with stoical, existentialist, or "ethical" advice. What his analyses show is that the patient often goes to the doctor so that constraints and precommitments may be imposed on him/her. This is why the appropriate question is indeed: "Why does one take care of oneself?" or "How is therapeutic rationality possible?" The doctors too often request a supplement of ethics from philosophy. However, philosophy does not have to intervene like wisdom or applied ethics; it is much more an instrument for analyzing and modeling situations including those related to the psychology and sociology of health.

Gérard Reach's book has all the merit of taking it seriously in this sense, including arguments and its own modes of conceptualization. This is why, in a single book, he produced a true work of philosophy and a major contribution to the psychology of health.

Pascal Engel
École des Hautes Études en Sciences Sociales

Preface and Acknowledgments

How can we accept that we ought to stop smoking, follow a diet, exercise, or take medications? The goal of this book is to describe the mechanisms of patients' adherence to long-term therapies, whose improvement, according to the World Health Organization, would be more beneficial than any biomedical progress. The fact is that the lack of adherence is a frequent phenomenon. For example, approximately half of the patients do not regularly follow medical prescriptions, resulting in deleterious effects on people's health and a strong impact on health expenditure.

This book, subtitled *Mind and Care*, describes how our beliefs, desires, and emotions intervene in our choices concerning our health. It investigates the moral rationality of adherence, by referring to concepts developed within the framework of the philosophy of mind. In particular, it tries to explain how we can choose between an immediate pleasure and a remote reward—preserving our health and our life. We postulate that such an "intertemporal" choice can be directed by a "principle of foresight" which leads us to decide to give priority to the future.

Just like patients' nonadherence to prescribed medications, doctors too often don't always do what they should: They are nonadherent to good practice guidelines. We propose that what was recently described as "clinical inertia" could also represent a case of myopia: From time to time doctors fail to consider the long-term interests of the patient. A chapter in this book is devoted to this issue; a complete analysis of nonadherence on the doctor's side, also published by Springer, can be found in a companion volume titled: *Clinical Inertia, A Critique of Medical Reason*.

Both patients' nonadherence and doctors' clinical inertia represent major barriers to the efficiency of care. If one thinks that overcoming these barriers would be beneficial, it is necessary to investigate their mechanisms, which is the scope of this book. However, it is also necessary to respect patients' autonomy. The analysis of the mental mechanisms of patient adherence, which is provided herein, sheds new light on the nature of the therapeutic alliance between doctor and patient. It is proposed that the dilemma between the principles of beneficence and autonomy can be analyzed in the framework of the relationship between mind and care.

This book was first published in French under the title: *Pourquoi se soigne-t-on, Enquête sur la rationalité morale de l'observance* (2nd edition, 2007), and I'm grateful to Dominique-Émile Blanchard, Jean-Luc Veyssy and Antoine Spire who accepted the publication of this book through their Edition House.

This English edition adds a number of new developments. I want to express my deep gratitude to Nastya Solovieva who translated the book, which was then edited by Jeff Engelhardt; and last but not least, to John Meyers who helped me through a number of criticisms and remarks to achieve the final version of the manuscript. Discussing chapter after chapter with John was most enjoyable.

I want to express my gratitude to Lisa M. Rasmussen for accepting the publication of this book in the Springer series "Philosophy and Medicine" and to Chris Wilby and Floor Oosting who organized at Springer the publication of this English edition. I would also like to extend my gratitude to Nathalie Huilleret for all her help in getting *Clinical Inertia, A Critique of Medical Reason* published.

Finally, I want to express my immense gratitude to Pascal Engel and Jon Elster. Pascal Engel's writings introduced me to the philosophy of mind: I discovered, with jubilation, another way of thinking about medicine, my patients' behaviors, and my own practice. He wrote the foreword to the French edition of this book, which is reproduced here. Jon Elster's writings helped me taste the emotional "alchemy of the mind" and appreciate the importance of Ulysses' myth for understanding the limits of our rationality. He wrote the foreword of the companion volume of this book, devoted to doctors' clinical inertia.

Contents

1 Introduction: The Doctor, Her Patient, and Their Reasons 1
1.1   Adherence and Nonadherence to Therapies: A Definition. 2
1.2   Nonadherence: How Common Is It? 2
1.3   The Consequences of Nonadherence 3
1.4   Scope of the Book 5
1.5   Some Simple Explanations for Nonadherence. 6
1.6   A Typology of Adherence? Analogous or Homologous
       Phenomena....................................... 7
1.7   The Real Question 9
1.8   From Behavior to Action 9
1.9   A Philosophical Understanding of Adherence
       to Long-Term Therapies.............................. 11
References... 12

2 The Classic View 15
2.1   Determinants of Nonadherence to Long-Term Therapies 16
       2.1.1   Intrinsic Factors 16
       2.1.2   Extrinsic Factors............................ 19
2.2   Behavioral Models of Patient Adherence................... 22
       2.2.1   The Health Belief Model 22
       2.2.2   The Theories of Reasoned Action
                and of Planned Behavior 23
       2.2.3   Theory of Interpersonal Behavior 24
       2.2.4   Leventhal's Self-Regulatory Model............... 24
       2.2.5   Transtheoretical Model of Change................ 25
       2.2.6   The Reversal Theory 27
2.3   Limitations of Psychological Models...................... 27
2.4   A New Perspective................................. 28
2.5   In Search of Mental Mechanisms in Psychology
       and Philosophy................................... 29
2.6   Observation, Explanation and Mechanisms................ 31

2.7   Patient and Agent   31
References   32

3 Intentionality   35
3.1   What Is 'In Your Head'   36
        3.1.1   The Different Types of Intentional Mental States   37
        3.1.2   The Place of Pleasure   39
        3.1.3   What Mental States Do   39
        3.1.4   Holistic Conception of the Mind   40
        3.1.5   The Background   41
3.2   A Mental Puzzle and Its Formation   42
        3.2.1   The Necessary Incompleteness of the Mental Puzzle ...   43
3.3   Actions   46
        3.3.1   Davidson's Causal Theory of Action   47
References   52

4 An Intentionalist Model of Patient Adherence   55
4.1   Therapeutic Agency   55
        4.1.1   To Take Care of Oneself or Not   57
4.2   An Intentionalist Model of Adherence   58
4.3   The Pivotal Role of Emotions in Patient Adherence   59
        4.3.1   Emotions, Boredom and Anxiety   61
        4.3.2   Emotions and Patient Adherence   62
4.4   Bringing Action into Play: Volition   64
References   66

5 The Dynamics of Intentionality   67
5.1   Motivational Force   67
5.2   Self-control   69
5.3   The Force of Habit   70
        5.3.1   Definition of Habit   71
        5.3.2   Mechanism of Habit   72
        5.3.3   Advantages of Habit   73
        5.3.4   Training Through Habit   75
        5.3.5   Adherence to Long-Term Therapies:
                  A Habit of Action   77
5.4   Intention, Decision, Resolution, and Willpower   79
        5.4.1   The Notions of Intention and Decision   79
5.5   The Dynamics of Intentionality   81
        5.5.1   To Take Care of Oneself Day After Day:
                  An Interpretation Within the Framework
                  of a Theory of Intentionality   81
        5.5.2   Back to the Mechanism of Habit   82
        5.5.3   Resolution and Willpower   83
References   86

6   Medical Irrationality 89
       6.1   Akrasia 91
       6.2   Patient Nonadherence to Therapy as a Case of Akrasia 92
              6.2.1   Philosophical Explanation of Akrasia 93
              6.2.2   A Choice Between Two Actions 93
              6.2.3   How Is Weakness of the Will Possible?
                       The Principle of Continence 95
              6.2.4   An Incomplete Explanation 96
              6.2.5   Second Explanation: The Partitioning of the Mind 96
              6.2.6   Partitioning of the Mind and Patient Nonadherence 99
       6.3   Another Medical Example of Irrationality:
              The Denial of Illness 100
              6.3.1   False Beliefs and Patient Nonadherence 103
       6.4   Logical Mechanisms of Irrationality 103
       References 105

7   Time and Adherence: A Principle of Foresight 107
       7.1   The Effect of Time 107
              7.1.1   Time and the Choice Between Two Desires 108
              7.1.2   Intertemporal Choice Between Two Rewards 111
              7.1.3   The Concept of Preference Reversal 111
              7.1.4   The First Solution: Precommitment Strategies 114
              7.1.5   Second Solution: Intermediate Rewards 116
              7.1.6   A Criticism of the Notion of Incontinent Action? 117
       7.2   The Principle of Foresight 119
              7.2.1   Temporality as a Criterion for Sorting the Content
                       of Mental States and the Principle of Foresight 120
              7.2.2   Implications of the Hypothesis 120
       7.3   The Appearance of Adherence 124
              7.3.1   From Animal to Human, a Phylogenesis of Patience 124
              7.3.2   Development of Patience in Children: Ontogenesis 125
              7.3.3   Neuroanatomy of Patience 125
              7.3.4   Neurobiology of Patience 126
              7.3.5   Genetics of Patience 127
              7.3.6   The Appearance of Belief 128
       7.4   A Pathophysiological Point of View 130
       7.5   A Top-down Model of Adherence 132
              7.5.1   Transmission of Principles 134
              7.5.2   Medicine and Health 135
       References 136

**8   An Intentionalist Account of Doctor-Patient Relationship
      and Biomedical Ethics.** 139
      8.1   Philosophical Analysis of the Doctor-Patient Relationship 140
      8.2   The Principle of Charity. 141
              8.2.1   Four Difficulties. 142
              8.2.2   Back to Patient Education 143
              8.2.3   Empathy 145
              8.2.4   Therapeutic Alliance 145
              8.2.5   Patient's Beliefs, Physician's Beliefs. 146
              8.2.6   The Therapeutic Relationship 148
      8.3   Adherence and Autonomy 149
              8.3.1   Therapeutic Autonomy in Medical Ethics:
                       Fourth or First Principle? 150
      8.4   Philosophical Conception of Autonomy
              as a Reflective Activity of the Mind. 151
              8.4.1   Reflective Activity of the Mind 151
              8.4.2   An Intentionalist Analysis of Autonomy 153
              8.4.3   Empirical Data: Patients Do Not Always
                       Wish to Exercise Their Autonomy. 154
              8.4.4   Therapeutic Autonomy and Models of the
                       Patient-Physician Relationship. 155
              8.4.5   Freely Giving up Autonomy. 158
              8.4.6   One's Own Physician: Healing One-Self. 159
              8.4.7   Theoretical Limits of Empowerment. 160
              8.4.8   Respecting Patient Autonomy 162
              8.4.9   Necessary Coexistence of Two Medical Models 162
              8.4.10   Training in Autonomy: For a Medicine of the Person ... 163
      References. 164

9   Doctors' Clinical Inertia as Myopia. 167
      9.1   Clinical Inertia: Definition and Logical Description 168
      9.2   Empirical Evidence: The Paradigm Case of
              Psychological Insulin Resistance. 170
      9.3   Empathy and Sympathy 172
      9.4   The Paradox of Empathy in Medical Care. 172
      9.5   Another Conception of Sympathy 173
      9.6   Care, Sympathy, Beneficence, and Love 174
      9.7   Care as a Special Form of Sympathy 174
      9.8   The Respective Values of Immediacy and Future 175
      9.9   Empathy, Sympathy, and the Ethical Dynamics
              of the Patient-Doctor Relationship. 176
      9.10   A Model of Chronic Care Involving Patient
              Education and Trust. 178
              9.10.1   Patient Education and Trust 181
      9.11   Conclusion: Mind and Care 183
      References. 183

10   Conclusion: Adherence Generalized 187
     10.1   A Choice Between Two Actions 187
     10.2   The Risk of Nonadherence............................... 188
     10.3   Generalization of the Problem 189
     10.4   Defining Adherence by Its Explanation 190
     10.5   Eros and Thanatos 191
            10.5.1   Why Do We Take Care of Ourselves?
                     The Two Meanings of Why 192
            10.5.2   Foresight, Prudence, and Happiness 193
            10.5.3   Eros 196
     References... 197

By the Same Author 199

Index 201

Chapter 1
Introduction: The Doctor, Her Patient, and Their Reasons

Abstract Patient nonadherence refers to a lack of coincidence between the patient's behavior and clinical prescriptions. At each step in the doctor-patient encounter—from making a first appointment, to undergoing screening tests, to taking medications and accepting changes in lifestyle, adherence is an issue: For instance, roughly half of the medication prescriptions are not filled. Nonadherence has been demonstrated repeatedly to erode the effectiveness of medical care and is linked with an increased rate in mortality. It has a major impact on health expenditures. A WHO report concluded that "increasing the effectiveness of adherence interventions may have a far greater impact on the health of the population than any improvement in specific medical treatment." In this book, I shall try to understand in general the phenomenon of nonadherence. To achieve this goal, I will attempt to describe what our patients are doing when they are adherent, for example, when they come to an office visit, take a tablet, stay on a diet or refuse a cigarette. These various manifestations of adherence must have something in common, i.e. their homology: My goal is precisely to discover what makes these phenomena homologous, without losing sight of differences. This will lead me to suggest that in each one of these cases we are dealing with not just a behavior, but an action. Thus I shall propose an interpretation of the mental mechanisms of adherence to long-term therapies based on the philosophy of human agency: Mind and Care.

A patient visits her doctor; the doctor makes a diagnosis, prescribes a medication, and the patient takes the medication as prescribed.

Experience shows that this is not always the case, by any means: A number of patients will never complete treatment for an acute illness, and the rate is even higher in chronic diseases. Our patient might not fill the prescription at the pharmacy, or stop the treatment prematurely, or follow only a portion of the doctor's recommendations. And when doctors themselves are patients, their compliance with prescribed treatment is no better, notwithstanding the fact that doctors are even less likely to have their own regular primary care physician and more likely to self-prescribe. The existence of physicians who are overweight, smoke cigarettes, do not exercise, or who abuse alcohol or drugs attests to the fact that sticking with treatment is not a problem limited to patients.

© Springer International Publishing Switzerland 2015

G. Reach, *The Mental Mechanisms of Patient Adherence to Long-Term Therapies*, Philosophy and Medicine 118, DOI 10.1007/978-3-319-12265-6_1

1.1 Adherence and Nonadherence to Therapies: A Definition

"Patient non-compliance" refers to a lack of "coincidence between the patient's behavior, in terms of taking medications, following diets, or executing lifestyle changes, and clinical prescriptions" (Haynes et al. 1979, 1–15). Aristotle understood this problem, as the epigraph of this book shows; it was not until the 1970s, however, that the term "compliance" became commonplace in medical parlance. This is perhaps partly in response to popular recognition of patient rights and growing awareness that medical science too is fallible. The diffusion of medical knowledge through the Internet has likely amplified this critical outlook. Today, the term "adherence" is preferred, as it suggests more active collaboration between physician and patient (Lutfey and Wishner 1999).

As this book will argue, adherence is not an all-or-none phenomenon, and varies not only between people, but also may vary in a given patient over the course of therapy. However, it is a general problem. At each step in the doctor-patient encounter—from making a first appointment, to undergoing screening tests, to taking medications, or any of the myriad other activities of modern healthcare, adherence is an issue.

1.2 Nonadherence: How Common Is It?

Bearing in mind the difficulty of knowing exactly which actions (or non-actions) are instances of nonadherence (for example, is not contacting your physician at the onset of an illness nonadherence?) rates of nonadherence are typically high. Roughly half of the medication prescriptions written in the United States are not filled: A study of 100,000 women taking an osteoporosis medication found that after 2 years, only 60 % of the total medication prescribed was actually taken (Curtis et al. 2009). Another review found that anywhere from 16 to 80 % of persons with diabetes do not stick with treatment over the long run (Cramer 2004). Yet another diabetes study found that two-thirds of patients followed dietary recommendations, but only one quarter adhered to advice on physical exercise. Only 7 % of the patients were adherent to all the treatment recommendations (McNabb 1997). Finally, when patients call the office to make their own appointments, 75 % will actually show up; but when the appointment is made on the patient's behalf (by a spouse, for example), the show-up rate drops to around 50 % (Meichenbaum and Turk 1987, 22).

Though these studies produced straightforward results, one should not get the impression that evaluating patient adherence is easy. It often depends on physicians' assessments, patients' self-observation, pill counts of untaken medication and, more recently, electronic surveillance systems that involve the placement of electronic circuits in the pill bottles registering each use (Blackwell 1997, 6).

Researching treatment adherence is complicated by the fact that it varies so widely: From the trivial (not taking a pill at the exact hour prescribed), to the catastrophic (going into a diabetic coma), to the "maybe serious, maybe not" (taking three of the four medications prescribed). Nonadherence can engulf the entire treatment, or be limited to one of its aspects. Moreover, adherence might vary over a period of time. A patient may be impressively adherent in the beginning of her treatment, but she may then suddenly become nonadherent; and later, just as suddenly, she may return to adherent behaviors. One would guess that this is a result of some events in her life—pregnancy is renowned for spurring a woman into adherence–but this isn't always the case. Often the reasons for patient behavior remain unavailable to the researcher, the treating physician, and even the patient herself.

It is unrealistic—and perhaps uncalled for—to expect perfect adherence. If a patient takes at least 80 % of a prescribed medication, for example, most practically-minded physicians would regard this as sufficient adherence. In this way, accommodation is made for patient forgetfulness, lapses in refilling a prescription at the pharmacy, and so forth. This forgiving approach also respects the fact that no system of safeguards can, in normal outpatient care, guarantee that the theoretical limit of adherence is met. However, with a disease like AIDS, it is very important that patients are 95 %-adherent: A lower rate runs the risk of inducing viral resistance.

However, a number of patients take fewer than 80 % of the prescribed pills: A study evaluated nonadherence in seven chronic diseases: Hypertension, hypothyroidism, type 2 diabetes, epilepsy, hypercholesterolemia, osteoporosis and gout. Sample sizes ranged from 4,984 patients for epilepsy to 457,395 for hypertension. Taking more than 80 % during the first year of therapy (good adherence) was observed in 72.3, 68.4, 65.4, 60.8, 54.6, 51.2 and 36.8 % of patients, respectively, for the seven disorders. Unexpectedly, the lowest adherence was observed in patients with gout, a disease in which flare-ups are renowned for their exquisite pain (Briesacher et al. 2008).

1.3 The Consequences of Nonadherence

Given this acknowledgement that routine medical care is able to tolerate some "slop", is it possible that nonadherence is not such a big deal after all? Perhaps–if patients took 80 % of their medication. But as we saw, the percentage is frequently much lower. Thus, unfortunately, nonadherence has been demonstrated repeatedly to erode the effectiveness of medical care. For example, in a study of antidiabetic medication use, researchers found that as adherence rates dropped, dangerously high blood sugars became more common (as measured by the percentage of glycated hemoglobin, or HbA1c) (Lawrence et al. 2006).

Nonadherence may have a direct impact on mortality. A study in the diabetes field showed that nonadherence is significantly associated with increased risks for all-cause mortality (Ho et al. 2006). The impact of nonadherence on mortality is quite strange, as shown by the following puzzling observation. In the Beta-Blocker Heart Attack Trial, the mortality at 1 year after a first myocardial infarction was higher in patients in the placebo group (3 %) than in the beta-blocker group (1.4 %). However, these rates were seen only among patients taking at least 75 % of the tablets (either the beta-blocker or the placebo). In nonadherent patients, the mortality rate in the beta-blocker group was 4.2 %—in other words, taking less than 75 % of the medication was worse than taking correctly the placebo! And even more intriguingly, among patients who were non-compliant with the placebo, the mortality rate was 7 % (Horwitz et al. 1990). This remarkable study demonstrates that adherence (whether to drug or placebo) is a substantial factor determining mortality. These curious findings have been replicated in a number of studies (Simpson et al. 2006).

Why should nonadherence to a placebo lead to the highest mortality rate? One explanation is that it is a reflection of a more general nonadherence to healthy behaviors. Nonadherers perhaps are less likely to follow a healthy lifestyle, with nonadherence to the medication (beta-blocker or placebo) being just one example. In support of this interpretation is a 2009 study which found that patients who were adherent with one medication (a cholesterol-lowering drug) were more likely to be adherent to a second medication (for osteoporosis) as well. In addition, the adherent patients were also more likely to follow through with screening tests such as mammograms and colonoscopies (Curtis et al. 2009). Recently, we observed that declaring that one does not fasten seatbelt when seated in the rear of a car is an independent determinant of nonadherence to medication in a validated questionnaire (Reach 2011).

Nonadherence can be financially costly as well, mostly through an increase in hospitalization (Lee et al. 2006; Sokol et al. 2005). For instance, in one reported case, a patient who skipped several doses of a diuretic medication (15 cents worth) was hospitalized for treatment of fluid overload. The six day hospital stay cost was $10,000 (Urquhart 1999, 119–145). In the United States, the economic cost of treatment nonadherence is estimated at $100 billion annually (Vermeire et al. 2005). There may be a vicious circle between nonadherence and associate health care costs (Iuga and McGuire 2014): Medication nonadherence leads to poor outcomes, which then increases health care service utilization and overall health care costs. The financial pressure is passed to patients by payers through higher copayments. Increased patient cost sharing beyond a threshold negatively impacts the level of medication adherence. An analysis of literature showed that patient cost sharing is associated with nonadherence (Eaddy et al. 2012).

A wide range of medical and public health areas are concerned with understanding and mitigating the effects of nonadherence: Management of AIDS, asthma, diabetes, hypertension, organ transplantation; schizophrenia and other serious mental illnesses; obesity and smoking; and even non-medical concerns, such as seatbelt use. Indeed, the effects of nonadherence are so pervasive that the World Health Organization noted in 2003 that "increasing the effectiveness of adherence interventions may have a far greater impact on the health of the population than any improvement in specific medical treatment (Sabaté, WHO report 2003)." What this means is that getting patients to adhere to existing treatments may be more important than discovering new treatments. New treatments avail us nothing if we don't actually use them.

Imagine a disease which causes 100,000 deaths per year, with a medication A that saves 20 % of patients, therefore 20,000 people. But if medication A is prescribed to only 80 % of patients which could benefit from it, it will save only 16,000 people. One would need a medication B saving 25 % of lives to have the same effect (to save 20,000 people) when it is given to 80 % of patients, as medication A if it were prescribed to everyone. Now, if medication A is prescribed to only 60 % of patients, medication B should save 33.3 % of patients: The greater the gap of lack of prescription, the more the increase in the effectiveness of medications to compensate for it becomes important, at a level which may be unrealistic. It should thus be more profitable to tackle the problem of access to care than to develop new medications (Woolf and Johnson 2005). The access to care includes patients' adherence to medication.

To date, efforts to improve treatment adherence have met with scant success: In a review of 83 adherence interventions reported in 70 randomized, controlled clinical trials, only 36 were associated with improvements in adherence and only 25 interventions led to improvement in treatment outcome (Haynes et al. 2008). This relative failure suggests that the medical and public health professions—and perhaps our society more generally—are missing something. The apparent inability to solve what seems to be a well identified problem is the motivation of this book.

1.4 Scope of the Book

In taking a step back to see the problem of nonadherence anew, we consider this question: How well do we understand adherence itself? Perhaps we fail to understand nonadherence *because we don't really understand adherence*. Why, after all, do some people take care of themselves in the first place?

This book investigates not only the how of adherence, but the why. Why, for example, does a patient take a blood pressure medication which has no discernible benefit and may have bothersome side effects? Why would the reformed smoker refuse a single cigarette, even though it will have no deleterious effect and will definitely provide pleasure? Why does a person take all of an antibiotic prescription when taking all but the last dose would be just as effective?

How indeed then does a person choose adherence (or nonadherence)? Certainly, we understand why the doctor makes his recommendations—that is, we know why the doctor wants the patient to be adherent—but why does the patient choose to follow (or not) those recommendations?

We may ask a more basic question: Is it *a choice*?

For example, how can we understand the curious behavior of the 20 % of transplant recipients who do not take their anti-rejection medication? (Rovelli et al.

1989) Can we decide between the physician's reasons and her patient's? And how is it possible that some people engage in such behaviors, where nonadherence seems to contravene one's own health interests?

In order to make headway on these questions, we must undertake a more global, perspective, and this perspective will necessarily be *philosophical*.

As a starting point, let us assume that people—doctors and patients—have their reasons for what they do. Let us set aside dismissive explanations such as "the patient is being irrational", or the even more unhelpful "she's being emotional". This starting assumption does not mean that every reason is clear, conscious, sensible, or consistent over time; we shall see that many reasons are opaque, transient, or unconscious, yet every bit as significant when it comes to understanding why people do what they do.

1.5 Some Simple Explanations for Nonadherence

Ignorance: If a patient does not understand what she needs to do, she cannot follow her doctor's recommendations. For example, some patients do not know how to use asthma inhalers unless instructed, and may administer the medication improperly. Some people believe that a seatbelt is not necessary when sitting in the back seat.

Forgetfulness: Patients forget to take medications, forget a doctor's appointment, forget to fast before blood drawing for cholesterol levels, and so forth.

Ignorance and forgetfulness, though pervasive, are usually easier to ameliorate: The use of educational brochures, teaching by specialized nurses (as in diabetes care, breastfeeding instruction, etc.), medication timers, and automated telephone appointment reminders are all innovations which have reduced ignorance and forgetfulness. But it is clear that nonadherence can also be intentional: Some patients very frankly say that they don't want to follow the advice they are given. After all, what would one say today of a patient who refused a bloodletting at the time of Molière?

On the other hand, some patients may believe that their doctor won't prescribe an antibiotic because the insurance company doesn't want him to; or that the doctor is ordering a test for defensive/legal reasons—or that the doctor is more worried than the patient, etc. In short, patients may think that doctors are also not fully autonomous, and react accordingly.[1]

Thus we may suppose from the start that two factors are at work. First, there is the patient's understanding of the prescription. The explanation of the prescription may have been insufficient, or the treatment may be so complex as to be virtually incomprehensible. *Health literacy* is defined as "the degree to which individuals have the capacity to obtain, process and understand basic health information and

---

[1] I am grateful to John Meyers for this remark.

services needed to make appropriate health decisions". Diabetic patients classified as having a low health literacy less frequently have a basic knowledge of diabetes care and more frequently have a high HbA1c level and retinopathy. *Health numeracy* refers to "the degree to which individuals have the capacity to access, process, interpret, communicate, and act on numerical, quantitative, graphical, biostatistical, and probabilistic health information needed to make effective health decisions". Patients with a low level of numeracy have a lower ability to perform a number of tasks required for their treatment, such as carbohydrate counting, identification of self-monitored blood glucose values within the target range and adjustment of insulin doses (Cavanaugh et al. 2008; Reach 2009). Or the patient might not grasp the importance of the advice. For example, consider packs of cigarettes bearing the warning "smoking can cause cardio-vascular disease". It is not certain that everyone understands what that means, and the warning "smoking can cause serious health problems" may mean very little to someone who has never been sick. In this case, we are not truly dealing with nonadherence, but with a failure of communication, a failure that a new medical field, patient education, is now trying to correct.

But as gratifying as it is to address fixable problems, the fact remains that nonadherence cannot be due to cognitive problems alone: The case of the overweight physician who smokes is proof enough that countless years of education and well-honed rationality are no match against the appetite for food and nicotine.

One might object—smoking is an *addiction*, and the smoker cannot stop smoking because of the symptoms of withdrawal, which appear as soon as she quits. And while this is an important factor in explaining the perpetuation of the habit, it does not explain why some smokers resume after months or years of abstinence. And what about other manifestations of nonadherence, in which one ignores medical prescriptions or advice concerning diet or physical exercise? Obviously, addiction is not a sufficient reason.

Clearly this first, simplistic explanation does not adequately explain patient nonadherence.

1.6 A Typology of Adherence? Analogous or Homologous Phenomena

Are some people more adherent, in general, than others? Is there some commonality shared by the endless variety of adherence behaviors, a quality which is stable and perhaps even measurable? The intuition is that a common denominator will help us understanding the phenomenon and will have heuristic value.

One starting point for delving into adherence phenomena more deeply is therefore an analysis of *analogy* and *homology*, like Roy Wise and Michael Bozarth did when they tried to set up a general theory of addiction (Wise and Bozarth 1987). They noted that "in biology, there are examples of superficially similar behaviors or organs that have evolved independently": For these "analogous" behaviors or organs look similar, but one cannot draw further conclusion from their similarity. They gave as examples the eye of the octopus and the eye of the vertebrate, the jealousy of the goose and the jealousy of the human: "In each case, the analogous details are striking, but there is no commonality of origin, and thus no *necessary* commonality of mechanism." By contrast, "homologous" organs or behaviors derive from common ancestral origin and, in biology, from common embryonic tissue, whereas analogies do not. Here "knowledge of one of a set of homologous organs or behaviors almost necessarily has some degree of heuristic value for the study of the others, even if the organs or behaviors are superficially dissimilar" and they gave as examples the wings of bats and birds, the fins of dolphins and whales, and the limbs of dogs and humans.

Human behaviors can be profitably organized along these lines: Elster, in his far-reaching work on social behavior, calls two behaviors homologous if they accomplish the same end; behaviors which involve the same physical actions but which have different intended outcomes are analogous (Elster and Skog 1999). For example[2] consider these behaviors:

1. Yelling "stop!" at a child running into the street.
2. Yelling "stop!" while playing a game with a child.
3. Grabbing a child's arm as he runs to the street

The first two behaviors, nearly identical in their outward features, are analogous behaviors in that they have the same external features. However, if we ask which behaviors are most similar in terms of their intention and underlying meaning, (1) and (3) are: Both are actions intended to keep a child from running into the roadway and getting hurt. These two behaviors have a homologous relationship.

Homology refers to functional similarity; analogy refers to structural similarity. Analogy helps us understand *how* something works; homology helps us understand *why*. Searching for homologies among diverse phenomena is a first step towards explaining those phenomena. For example, knowledge of the reproduction or the metabolism of whales can help us form hypotheses about bats, and vice versa. This is why, as pointed out by Wise and Bozarth, discovering a homology has a heuristic value: In the case of homologous phenomena, their definition becomes *ipso facto* inseparable from their explanation.

This will be precisely the method used in this book: I will try to *explain* the phenomenon of nonadherence which, *by definition*, is opposed to the effective completion of a medical treatment and can manifest itself in any stage of the treatment. As we have seen, it is clearly not the same thing to smoke or to omit taking one's pills, and these two behaviors cannot be treated (in the medical sense of the word) in the same way; and yet, they must have something in common. My goal is precisely to discover what makes these phenomena homologous and not simply analogous, without loosing sight of differences. I believe that it is only by

---

[2] An illustration given by John Meyers.

following these steps that we may hope to explain nonadherence, *to understand it in general.* And understanding it in general is the object of this book.

1.7 The Real Question

Nonadherence seems irrational, it makes no sense. Why would someone not take a prescribed medication after going to the trouble of visiting the doctor in the first place? To keep one's health, to avoid putting one's life at risk, aren't these the goals of the reasonable person? Shouldn't we then conclude that those who do not are irrational?

The doctor who has to deal with a nonadherent patient is often amazed and even exasperated. But, as we have said, the patient who doesn't take her pills must surely have a reason—when can we really say that those reasons aren't good enough? Who is to decide between the doctor's reasons and the patient's reasons if they should differ? The medical profession has a great deal to say about *how* one might take care of a medical problem. The problem of nonadherence forces us to address *why* one might take care of a medical problem: *Why do we take care of ourselves at all?*

Nonadherence perplexes the physician because it involves two paradoxes: First, it is both rational and irrational. Its rational to not take a medication which has no near-term benefit, yet its irrational to miss the long-term benefits. Likely, its irrational to drive rather than fly (as many did after the 9/11 terrorist attack), yet its rational to choose a mode of travel which allows for more control if problems start to arise (as being the driver of a car does, but not being a passenger on an airplane).

There is another paradox, maybe more subtle: Nonadherence is both *natural* and irrational. As we shall see, our reasons for doing something depend critically on how we see our future, and how far into that future we look. We will see that some of us are unable to look far into the future, making it natural (and therefore rational!) to be nonadherent. Yet sometimes we also feel that such a behavior is irrational, since we *know* that we are acting against our own interest.

1.8 From Behavior to Action

Patient adherence and nonadherence are behaviors, and, as such, are the proper study of psychology. There is a wealth of literature in this discipline concerning the matter of adherence to therapies. Psychological methodology is varied, but its essence consists of: (1) observing human behavior, (2) modeling it, and (3) testing the models. One such model developed by psychologists is the *Health Belief Model*, which will be described in more detail shortly.

But it is also possible to view the problem from a different angle, and it is this angle that we shall focus on. Adherence and nonadherence are two sides of a coin which embody a deep paradox in human nature. Beyond the phenomena themselves, beyond traditional psychological explanations of causation, lies a *philosophical* question which holds the key to this vexing clinical problem.

The philosopher of mind Pascal Engel, commenting on a Somerset Maugham novel—in which an overweight woman goes on a diet but then stuffs herself more than ever—points out the direction our inquiry will take:

> The writer is interested in [nonadherent persons] because she wants to show a particular trait of human nature, the psychologist because she wants to know how these things happen. The philosopher wonders how these things are *possible* (Engel 1991).

David Pears similarly describes the difference between psychology and philosophy: Philosophers are interested in the conceptual line that separates the possible from the impossible. The psychologists want to see how certain phenomena exist: Their question is not: 'how can these things happen', but rather 'how do these things happen' (Pears 1998, 1).

In this book, I will attempt to describe what our patients *are doing* when, for example, they come to an office visit, take a tablet, stay on a diet or refuse a cigarette. This will lead me to suggest that in each one of these cases we are dealing with not just a behavior, but *an action*: Actions encompass behaviors and all their associated underpinnings (meaning, intention, etc.). In moving from the study of behavior to the study of action, we necessarily move beyond the traditional bounds of psychology into the realm of philosophy.

Patient nonadherence, as will be shown, may be far better understood from this *action* perspective: It is an instance of *incontinent* action. We perform an incontinent action when we do something even though we know that, all things considered, we shouldn't be doing it. The concept of incontinence has been used by philosophers since at least the time of Aristotle, and modern philosophers have drawn many illuminating insights from this puzzling phenomenon.

We will see that by applying some of these insights we will come to better understand patient nonadherence. One central idea is the *principle of foresight*, which will be defined and elaborated in this book. We will find that patient adherence and nonadherence are outward expressions of the presence or absence of a deeper faculty, that of foresight.

Deep down, the problem is to understand how we choose between options which often differ in their temporal aspect: Nonadherence is usually satisfying in the concrete, here-and-now, while adherence aims at a necessarily more distant and abstract reward, such as lengthening one's life or reducing the chances of developing emphysema. The study investigating adherence in seven chronic diseases, quoted above, found that young age was a strong predictor of nonadherence in six of them (Briesacher et al. 2008). It is tempting to explain this finding by hypothesizing that in chronic diseases, the choice between a smaller-sooner, and a larger-later, reward will have to be made day after day on a longer term basis in younger patients, increasing the risk of non persistence to therapy. This problem of "intertemporal choice" is currently the object of numerous studies and it will be at the heart of our investigation.

1.9 A Philosophical Understanding of Adherence to Long-Term Therapies

So far, we have been using terms such as "belief", "intention", "desire", and "choice" in an open-handed and naïve way. But as we search for less casual, more precise definitions of these everyday ideas to better understand what role they play in generating our actions, philosophy of mind again comes to our aid. Frank Ramsey, the British mathematician and philosopher, noted in his essay entitled *Philosophy* (1929) that

> In philosophy we take the propositions we make in science and everyday life, and try to exhibit them in a logical system with primitive terms and definitions, etc (Ramsey 1990).

Similarly, we shall try to craft a logical framework of simpler concepts to help us understand the how and why of human action, and therefore of adherence to medical treatment.

Analytic philosophy, or more generally, *philosophy of mind*, attempts to describe the mechanisms which connect 'mental states', such as knowledge, skills, beliefs, emotions, desires, and even visceral perceptions (for instance, hunger), using logically primitive terms and concepts. Our goal is to understand what we mean when we talk about the 'reason' for a behavior (for example, why I do or don't take my medication), by asking the question: In general, *why do I do this?*

This book proposes a philosophical interpretation of the problem of adherence to long-term therapies. Our interpretation leads to a theoretical model in which mental states interact in a hierarchical manner, and in which emotions and desires, rather than beliefs, have priority—in contrast to the cognitive emphasis in classic psychological models. Thus one of the ambitions of this work is to show how the application of philosophical concepts sheds new light on issues in medical anthropology (non adherence, disease denial, the doctor-patient relationship); and how in turn it may enrich philosophical concepts with empirical medical research.

In the beginning of this introduction, we saw that when the doctor writes a prescription and when the patient follows or doesn't follow the medical advice, both have *their reasons* for doing so. Applying concepts from philosophy of mind to the domain of medical anthropology, we will find a new theoretical basis for the relationship between doctor and patient. We may describe it as a relationship between their reasons. The reasons of care: *Mind and Care.*

Following this first introductory chapter, Chap. 2 is an overview of classical psychological models of nonadherence. Chapter 3 introduces basic philosophical concepts, and presents a short account of the concept of "Intentionality". Chapter 4 provides an "intentionalist" model of adherence. Chapter 5 presents a dynamic view of intentionality, by integrating in this model the concepts of motivational force, self-control, habit and resolution. Chapter 6 describes patient nonadherence as a case of weakness of will, or *akrasia*. Chapter 7 considers more specifically the temporal dimension of adherence and nonadherence in chronic diseases, focusing on the description of a principle of foresight, a concept introduced in this book: Nonadherence may be understood as a failure to give priority to the future. Chapter 8 outlines the consequences of this insight on the therapeutic alliance between doctor and patient and addresses ethical issues. Chapter 9 shows that doctors too may fail to consider the future interests of the patient: Thus, like patients' nonadherence to medical recommendations, doctors' *clinical inertia* could represent a case of *clinical myopia*. Chapter 10 generalizes the problem of adherence and proposes a relationship between the fact of taking care of oneself and self-love.

References

Blackwell B, editor. Treatment compliance and the therapeutic alliance. Amsterdam: Harwood Academic Publishers; 1997.

Briesacher BA, Andrade SE, Fouayzi H, Chan KA. Comparison of drug adherence rates among patients with seven different medical conditions. Pharmacotherapy. 2008;28:437–43.

Cavanaugh K, Huizinga MM, Wallston KA, Gebretsadik T, Shintani A, Davis D, Gregory RP, Fuchs L, Malone R, Cherrington A, Pignone M, DeWalt DA, Elasy TA, Rothman RL. Association of numeracy and diabetes control. Ann Intern Med. 2008;148:737–46.

Cramer JA. A systematic review of adherence with medication for diabetes. Diabetes Care. 2004;27:1218–24.

Curtis JR, Xi J, Westfall AO, Cheng H, Lyles K, Saag KG, Delzell E. Improving the prediction of medication compliance: the example of bisphosphonates for osteoporosis. Med Care. 2009;47:334–41.

Eaddy MT, Cook CL, O'Day K, Burch SP, Cantrell CR. How patient cost-sharing trends affect adherence and outcomes: a literature review. PT. 2012;37:45–55.

Elster J, Skog O-J, editors. Getting hooked, rationality and addictions. Cambridge: Cambridge University Press; 1999.

Engel P. Foreword to his translation of Donald Davidson's paradoxes of irrationality, Éditions L'Éclat; 1991.

Haynes RB, Taylor DW, Sackett DL. Compliance in health care. Baltimore: Johns Hopkins University Press; 1979.

Haynes RB, Ackloo E, Sahota N, McDonald HP, Yao X. Interventions for enhancing medication adherence. Cochrane Database Syst Rev. 2008;2(2):CD000011.

Ho PM, Rumsfeld JS, Masoudi FA, McClure DL, Plomondon ME, Steiner JF, Magid DJ. Effect of medication nonadherence on hospitalization and mortality among patients with diabetes mellitus. Arch Intern Med. 2006;166:1836–41.

Horwitz RI, Viscoli CM, Berkman L, Donaldson RM, Horwitz SM, Murray CJ, Ransohoff DF, Sindelar J. Treatment adherence and risk of death after myocardial infarction. Lancet. 1990;336:542–5.

Iuga AO, McGuire MJ. Adherence and health costs. Risk Manag Healthc Policy. 2014;7:35–44.

Lawrence DB, Ragucci KR, Long LB, Parris BS, Helfer LA. Relationship of oral antihyperglycemic (sulfonylurea or metformin) medication adherence and hemoglobin A1c goal attainment for HMO patients enrolled in a diabetes disease management program. J Manag Care Pharm. 2006;12:466–71.

Lee WC, Balu S, Cobden D, Joshi AV, Pashos CL. Prevalence and economic consequences of medication adherence in diabetes: a systematic literature review. Manag Care Interface. 2006;19:31–4.

Lutfey KE, Wishner WJ. Beyond "compliance" is "adherence", improving the prospect of diabetes care. Diabetes Care. 1999;22:635–9.

McNabb WL. Adherence in diabetes: can we define it and can we measure it? Diabetes Care. 1997;20:216–8.

Meichenbaum D, Turk DC. Factors affecting adherence. In: Facilitating treatment adherence. New York: Plenum Press; 1987.

Pears D. Motivated irrationality. South Bend: St Augustine's Press; 1998.

Ramsey FP. In: Mellor DH (ed) Philosophical papers. Cambridge: Cambridge University Press; 1990.

Reach G. Linguistic barriers in diabetes care. Diabetologia. 2009;52:1461–3.

Reach G. Obedience and motivation as mechanisms for adherence to medication. A study in obese type 2 diabetic patients. Patient Prefer Adherence. 2011;5:523–31.

Rovelli M, Palmeri D, Vossler E, Bartus S, Hull D, Schweizer R. Compliance in organ transplant recipients. Transplant Proc. 1989;21:33–844.

Sabaté E (ed). World Health Organization report. Adherence to long-term therapies, evidence for action, Geneva, Switzerland; 2003.

Simpson SH, Eurich DT, Majumdar SR, Padwal RS, Tsuyuki RT, Varney J, Johnson JA. A meta-analysis of the association between adherence to drug therapy and mortality. BMJ. 2006;333:15.

Sokol MC, McGuigan KA, Verbrugge RR, Epstein RS. Impact of medication adherence on hospitalization risk and healthcare cost. Med Care. 2005;43:521–30.

Urquhart J. Pharmacoeconomic impact of variable compliance. In: Metry JM, Meyer UA, editors. Drug regimen compliance: issues in clinical trials and patient management. Chichester: Wiley; 1999.

Vermeire E, Wens J, Van Royen P, Biot Y, Hearnshaw H, Lindenmeyer A. Interventions for improving adherence to treatment recommendations in people with type 2 diabetes mellitus. Cochrane Database Syst Rev. 2005;18:CD003638.

Wise RA, Bozarth MA. A psychomotor stimulant theory of addiction. Psychol Rev. 1987;94:469–92.

Woolf SH, Johnson RE. The break-even point: when medical advances are less important than improving the fidelity with which they are delivered. Ann Fam Med. 2005;3:545–52.

Chapter 2
The Classic View

Abstract A purely descriptive analysis of factors playing a role in patient adherence, related to the patient, her disease and the health care system, will not easily reveal the underlying psychodynamic processes that shape a patient's adherence. To go further, various behavioral models have been proposed to put these different factors into a conceptual framework that accounts for their interactive production of adherent or nonadherent behavior: The Health Belief Model, the Theories of Reasoned Action and of Planed Behavior, the Theory of Interpersonal Behavior, the Self-Regulatory Model, the Transtheoretical Model of Change and the Reversal Theory will be briefly described in this chapter. These models demonstrate statistical correlations between mental states and certain behaviors; but, as is well known, statistical correlations do not imply causal relations. Psychological models have therefore a major limitation: They cannot explain why an individual is or is not adherent to the medical advice that she is given. At the level of the individual, specific behaviors remain wholly unexplained and mysterious. In other words, behavior averaged out over a population is no longer sufficient for our investigation; rather, I shall focus on what a given patient is actually doing and why (i.e. for what reasons) she is doing it. I seek to establish a theory, taking its roots in the philosophy of mind, that defines what is meant by the 'reasons of care' and which shows how these reasons bring about caretaking—therapeutic—actions, supporting a causal relationship between Mind and Care.

Nonadherence to medical treatment has initially been attributed to ignorance on the part of the patient. Nowadays, it is understood that this explanation falls well short of capturing the many reasons why treatment doesn't happen as planned. If simple ignorance were the only reason for treatment nonadherence, then patient education would eliminate the problem. Both clinical experience and research suggest that knowledge is not enough. For example, Meichenbaum and Turk have shown that there is little correlation between the extent of a patient's knowledge of disease and adherence to her treatment (Meichenbaum and Turk 1987, 61). This is not to say that patient information has no role to play: Public health measures to encourage hand washing, safe sex, and helmet use have all proven successful in increasing healthful behaviors. It is also the case that for some patients, more knowledge leads to improved adherence, while for other patients it has little effect.

© Springer International Publishing Switzerland 2015

G. Reach, *The Mental Mechanisms of Patient Adherence to Long-Term Therapies*,
Philosophy and Medicine 118, DOI 10.1007/978-3-319-12265-6_2

Patient information is essential in the care of chronic disease; however, it is not sufficient to lead to patient adherence. Informing patients is one thing, but motivating them is yet another.

We may divide the factors contributing to nonadherence into two categories: Those that are *intrinsic* to the patient—her knowledge based on the explanations she has received, her beliefs, and so on—and those *extrinsic* elements related to the disease and its treatment, the patient's economic and social status, and other features of the local environment.

2.1 Determinants of Nonadherence to Long-Term Therapies

2.1.1 Intrinsic Factors

2.1.1.1 Lack of Knowledge

Knowledge of one's illness is an important—though neither necessary nor sufficient—factor determining adherence. Similarly, knowledge regarding the treatment is important, but in ways often overlooked by the physician. The crux of diabetes treatment, for instance, is maintaining proper insulin levels in the patient's body through insulin dose adjustment. And while this is true from a pathophysiological point of view—the dominant view in modern medicine—it is not sufficient to bring about a successful treatment. An endless number of distinct actions and decisions are needed to enact the seemingly simple concept of "maintaining proper insulin levels." Each facet of treatment is a point of potential breakdown: It is an unwise doctor indeed who brushes aside these "trivial" matters. And yet, these issues have little to do with pathophysiology per se.

Communication is yet another concern in providing education to the patient and in implementing treatment. In the United States, for example, around 25 % of physicians are from other countries, and many Americans do not speak English as their mother tongue. The chances are high that a doctor and her patient will have some communication difficulties simply on this basis. However, a linguistic barrier may exist even when the patient and her doctor speak the same language (Reach 2009). In addition, there is the oft-criticized tendency of physicians to speak too quickly and using too much jargon. Patients complain of inadequate opportunity to ask questions, or even time to formulate questions before their physician has breezed out of the exam room.

2.1.1.2 Wrong Beliefs

The beliefs which a patient brings to the physician's office are vital in determining the course of the treatment. How does the patient envision her illness, her treatment, her vulnerability, her capacity to take care of herself, and the power of medicine to make a difference in the progress of the disease? Meichenbaum and Turk give some particularly revealing examples of the beliefs that can lead to nonadherence to treatment. To mention just a few: 'If you take the medication too often, you can develop a resistance to it', or 'you become addicted to it', or 'the medication doesn't do anything', 'they are trying to poison me', 'God will cure me of the disease', 'complications only happen to others', 'how will I know that I don't need the medication anymore if I continue to take it' or 'nothing works for me', etc. (Meichenbaum and Turk 1987, 47). These facts are not necessarily derived from the doctor's explanations and might not even be conscious (Laplantine 1997, 246–265). They typically come from the patient's family, culture, or ethnic origin, and the patient's prior experiences: Competing ideas about an illness and its treatment may be found in books, magazines, on television, and the Internet, and patient beliefs are shaped by all these sources and then some.

2.1.1.3 Biases

Amos Tversky and Daniel Kahneman observed that people, when they have to make decisions in a context of uncertainty, usually do not behave according to the predictions of the classical Expected Utility Theory, where one makes choice on the basis of the value (the "utility") of the outcomes and their respective probabilities. Instead, they use "heuristics": These are simple and efficient rules which work well under most circumstances but which in certain cases lead to systematic errors or cognitive biases. One of these heuristics is known as the availability heuristic (Tversky and Kahneman 1974): Suppose you are asked to evaluate the relative frequency of cocaine use in Hollywood actors, you may assess how easy it is to retrieve examples of celebrity drug-users (Gilovich and Griffin 2002, 1–18). Using this heuristic may obviously introduce a bias in the estimation.

This effect may be relevant in our context. For instance, suppose a patient trying to evaluate the relative risk of hypoglycemia after increasing the dose of insulin. She may do it by assessing how easy it is to retrieve examples of hypo in her past experience. Since we usually remember more readily unpleasant events (Baumeister et al. 2001), she will overestimate the risk. Accordingly, she will not do what she was taught: To increase the dose when blood glucose is high.

In addition, the Prospect Theory proposed by Kahneman and Tversky suggests that in our mind, a loss of X \$ is more averse than a gain of X \$ is attractive (Kahneman and Tversky 2000, 1–16).This loss aversion may also be relevant for the issue of insulin dose adjustment: Just consider the risk aversion effect on weighing the risk of hypoglycemia (loss) versus the gain linked to getting a better blood glucose (Reach 2013).

2.1.1.4 The Effect of Uncertainty

Patients may not follow clinical advice because of awareness that this advice is ill-funded, or even may change over time. Interestingly, Anderson, in a review entitled: *"The Psychology of Doing Nothing"* (Anderson 2003), proposed that a major reason for privileging status quo (nonadherence—not taking the pill is a form of status quo) is the difficulty of the choice. Among the many factors which can result in making difficult a decision and to lead to the inaction *in general*, one can retain the following: The difficulty in adopting a clear strategy by lack of time, the multiplicity of the options, uncertainty on the preferences, the fact that the choice is badly defined, perhaps the personality of the agent, even her culture.

2.1.1.5 Emotions

Emotional states affect treatment adherence. Depressed persons are more likely to judge treatment to be pointless or even undeserved. Highly anxious persons may avoid going to the doctor for fear of what bad news they may receive. Grandiose or euphoric patients may stop treatment as soon as they please; some emotionally disturbed patients may sabotage their own treatment as a way to rebel against their physician.

Some patients are especially sensitive to losing control. The onset of a new illness, and a doctor's orders for its treatment, may set in motion an instinctive "push back"—what psychologists call "reactance"—which can lead to nonadherence (Brehm 1966). Persons who see themselves as freely making their own choices in life may rebel against any infringement upon this freedom. Indeed, presenting a medical prescription in an authoritative way was shown to lead to patient reactance and nonadherence (Fogarty 1997; Fogarty and Youngs 2000).

Interestingly, this prideful, "I'll be damned" retort may happen even in the face of full knowledge of the health consequences. Some smokers, for example, see smoking as a measure of their independence and freedom. Cigarette advertisements often highlight this: The Marlboro man on the open plain, the liberated woman smoker ("you've come a long way, baby."). Smoking by teens is sometime nothing but a way to show their independence or even rebellion. One may also note that the recent social debate about ending public smoking was often centered around issues of liberty and freedom.[1] Thus, some smokers fell that they control events rather than being subjected to them: For them, the important moment is the lighting of the cigarette (Elster and Skog 1999, 14).

By contrast, our recent observation that the behavior of fastening seatbelt when seated in the back of a car is more frequent in adherers to medication may be explained by the fact that some patients are adherent simply because they are, in general, obedient (Reach 2011a). This idea is consistent with the typological distinction between "critical" and "traditional" adherers proposed by Bader et al. for people living with AIDS, in which traditional ("unquestioning") adherers have the ability and willingness to follow a therapeutic regimen exactly as prescribed by a medical authority, based on a traditional, asymmetric doctor-patient relationship (paternalistic model). Among "traditional" adherers, Bader et al. described a

---

[1] I am grateful to John Meyers for these remarks.

subtype of "faithful" patients who are "obedient and yield readily in a subservient way to doctors' orders (Bader et al. 2006)."

In other cases and probably less consciously, some patients don't take their medication because it reminds them too much of their illness, and that's precisely what they would like to forget.

2.1.1.6 The Patient's Interpersonal World

Social/group forces in treatment adherence may be formidable. Young persons, in striving to fit into their peer group, may eschew treatment if they fear it will lead to ostracism. Risk taking behavior is typically increased by the presence of other people, and risk-taking and treatment nonadherence go hand-in-hand. Conversely, social isolation can reduce treatment adherence if this means less support and encouragement for the patient. There may be difficulties with one's children, lack of time, limited resources, the loss of a job, the breakup of a relationship, or social deprivation, to name just a few (Daley and Zuckoff 1999, 25).

2.1.1.7 The Patient, Her Doctor and Medicine

Of course, the physician's qualities are important in treatment adherence. A proper match between a physician's approach and the patient is critical in establishing and sustaining a productive doctor-patient relationship. Meichenbaum and Turk give a list of factors associated with adherence: The perception by the patient of the friendly and open character of the physician. Is she treated with respect and dignity? Does she participate in the decisions, does the physician take into account her expectations? Does the physician pay attention to her particular case, giving explanations to motivate her? Patient satisfaction with the physician and the treatment regimen is an important correlate of adherence (Meichenbaum and Turk 1987, 63–64).

It is perhaps most distressing of all to admit that the healthcare system itself can contribute to nonadherence. Beyond physician-patient relations, there is the very organization of the health care system: How easy it is to get an appointment, the quality of the reception, the frequency of the appointments. Subtler factors are at play: The coherence of what all the different members of the medical team are saying, the continuity of treatment, whether the patient is treated by the same doctor from one visit to the next. Hospital care can be very distressing in this respect; a patient might feel that she is being treated by an anonymous group rather than by "her doctor".

2.1.2 Extrinsic Factors

2.1.2.1 The Patient with a Silent Disease

Treatment for troublesome symptoms is usually adhered to more assiduously: The patient has an immediate and pressing motivation to stick with the prescribed treatment. Indeed, sometimes this leads to nonadherence of another sort: The patient may take more of the medication than prescribed, or use non-prescribed medications in conjunction with those prescribed. However, this is not always true. Consider the example of gout: Treatment aims to decrease uric acid levels below 6 mg/dl, and is only given after the patient has experienced a gout flare-up, which typically is exquisitely painful. Surprisingly, it is the chronic disease where adherence is the worst (Briesacher et al. 2008; Reach 2011b). Even more surprisingly, a recent study showed that use of non-steroidal anti-inflammatory drugs in the year prior to urate-lowering drug initiation (suggesting the occurrence of a crisis) was a significant predictor of poor adherence to subsequent gout therapy (Harrold et al. 2009).

Illnesses which are asymptomatic—including hypertension, diabetes, high cholesterol, and many cases of cardiac disease—present difficulties which bedevil treatment adherence. Medications may not be taken on schedule, the prescription not refilled, check-ups skipped, lifestyle modifications not made. It is ironic that modern medicine has fostered this problem by having so many effective, symptom-eliminating treatments available. Hypothyroidism, for example, was once a cause of much morbidity and mortality, but has been transformed by modern treatment to a silent disorder with few or no symptoms. No wonder that nonadherence sometimes occurs in hypothyroidism (Briesacher et al. 2008). This was referred to as "levothyroxine pseudo-malabsorption" (Ain et al. 1991).

2.1.2.2 Chronic Diseases: The Patient and Time

In general, treatment of chronic diseases is beset by more adherence issues than is treatment for acute diseases. Nevertheless, even treatment for acute illness is often stopped prematurely—as for example when patients do not finish a course of antibiotics—or adhered to selectively (taking the painkillers but not doing the exercises). Sometimes, a chronic illness is not recognized as such by the patient, who imagines that once a particular crisis is resolved, no more treatment is warranted. In the same vein, in the treatment of hypertension, the patient may wonder: Why should I continue to take the medication which lowers my blood pressure if the blood pressure is now normal? Some patients might interrupt the treatment either voluntarily or unconsciously, in an attempt to verify what she has been told: That the treatment must continue for the rest of her life.

Thus, a common thread runs through each of these common impediments to treatment adherence: An inability to maintain a sustained vision of treatment in the long-term. When immediate concerns routinely overwhelm longer term goals, incontinence—and treatment nonadherence—may be the result. The initial enthusiasm to "fight" an illness—or the excitement of a "flight into health"—typically wanes with time. As these energetic but short-lived emotions wane, so does treatment adherence.

A large part of this book will be devoted to this aspect of the problem. Treatment duration is a major extrinsic factor which bears upon adherence. In general, patients tend to drop out the longer the treatment. Briesacher and his

Fig. 2.1 Nonadherence is more frequent in younger people, originally published in Briesacher et al. (2008). Modified with kind permission of © Wiley 2008 and of the author. All Rights Reserved

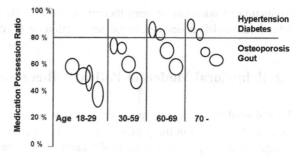

colleagues observed that an index of adherence, the Medication Possession Ratio (the ratio of the total days supply of medication that was dispensed divided by the number of days of the evaluation period) was lower in younger people (Fig. 2.1) (Briesacher et al. 2008).

This effect of duration can also be observed over a short period: For example, a study has shown that adherence to iron supplements progressively diminishes over the course of the three trimesters of pregnancy (Meichenbaum and Turk 1987, 60). However, as will be discussed later, there are important and fascinating reasons why *some patients* are better with long-term treatment than short-term. Clearly, this effect of duration will be the key to understand adherence.

2.1.2.3 Hic et Nunc: The Powerful Temptations of Advertising

We are surrounded by a quintessential extrinsic factor: Advertising. Much advertising is tailored to encourage us to buy and consume now rather than later, and advertising certainly works, at least from the seller's point of view. We are then confronted to a choice between a temptation, offered by ads, which is immediate and concrete, and the desire to remain healthy, which is remote and abstract.

Consumers are not passive in the purchasing/consuming process, of course; but by the same token, "caveat emptor" hardly scratches the surface of the complexity of the psychology of advertising, even if we adjudge it adequate in the legal arena. Advertisements for cigarettes, though now much curtailed, continue to have influence on certain target populations, such as teens. Alcohol is heavily advertised; television ads for hard liquors can now be seen on some cable channels. Vending machines for junk food may be found in schools and medical clinics. Stairways in public buildings are often hidden away, dissuading people from using them for even this modest bit of exercise. There is a direct correlation between the surge in obesity during the last 40 years and the number of cars per household as well as the number of hours spent watching television per week.

As we can see, the list of factors playing a role in patient adherence is long. But a purely descriptive analysis will not easily reveal the underlying psychodynamic processes that shape a patient's adherence. To go further, we must put these different factors into a conceptual framework that accounts for their interactive production of adherent or nonadherent behavior. This is the goal of the various behavioral models delineated in the next section of this chapter.

2.2 Behavioral Models of Patient Adherence

Several models attempting to understand how a health behavior can be changed have been proposed in the psychological literature. These models were often constructed at the request of public health authorities to help increase the efficacy of measures such as screenings for tuberculosis or anti-smoking campaigns. Given this, it should be unsurprising that their value is primarily statistical. As a science, health psychology strives to find statistically significant correlations between health behaviors and their putative determinants through rigorous 'empirical' research involving observable data–for instance, individuals' answers on questionnaires, findings on physical exam, results of lab tests, and the like. If the methods of information collection and the studied population are defined rigorously enough, the results of the research can be reproduced, demonstrating all the qualities of 'scientific' research, where the results do not depend on the investigator.

Let us consider a few of these models: The Health Belief Model, the Theories of Reasoned Action and of Planed Behavior, the Theory of Interpersonal Behavior, Leventhal's Self-Regulatory Model, the Transtheoretical Model of Change and the Reversal Theory.

2.2.1 The Health Belief Model

The first model to include cognitive factors in the determination of behavior was the *Health Belief Model*, developed in the early 1950s by Godfrey Hochbaum, Stephen Kegels and Irwin Rosenstock (Becker and Maiman 1975).

This model superimposes the perception of threats and expectations onto a socio-demographic background, which includes, for example, age, gender, ethnicity, profession, etc. To make the decision to adopt a new health behavior the agent must feel personally vulnerable, regardless of what the "objective" situation might be. Threats include the perception of the individual's own vulnerability in the face of a health problem and her perception of the problem's severity. The model considers severity not only in terms of health, (including pain, discomfort, and the risk of death), but also as regards its professional, social and family consequences. The expectations are the benefits that the individual anticipates from the health behavior, the individual's perception of her capacity to perform the action (self-efficacy), and her perception of the obstacles to performing it. Once the individual, having weighed the pros and the cons, has decided to submit to treatment, a cue might be necessary to trigger its implementation. This might be an internal event (the appearance of the first symptom, for example) or an external one (a media campaign or the loss of a relative to the same illness) (Fig. 2.2).

Fig. 2.2 The Health
Belief Model and patient
adherence. Modified from a
figure published in Janz and
Becker (1984). Reprinted
by permission of SAGE
Publications

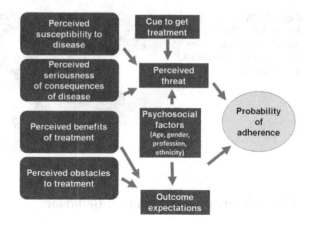

2.2.2 The Theories of Reasoned Action and of Planned Behavior

The Theory of Reasoned Action was developed in 1967 by Martin Fishbein and Icek Ajzen (Fishbein and Ajzen 1975). Its title implies that individuals are rational beings who use the information at their disposal and consider the consequences of their actions before performing them. This theory maintains that the behavior depends essentially on the intention of the subject to perform it. Here intention is described as the indication of the strength of the subject's desire to perform the behavior and the efforts that she plans to invest in order to reach this goal.

The intention of the individual to perform a particular behavior depends on two types of factors. The first factor is the individual's attitude towards the behavior, consisting of the positive or negative evaluation of the behavior. The attitude in turn depends on different beliefs of the patient concerning the consequences, positive or negative, of adopting the behavior. The second type of factor are the subjective norms, or the beliefs concerning the way the behavior is perceived by the people important to the patient (for instance, family, friends, the physician, the police) and her more or less intense desire to follow their advice. Ajzen later modified this model by another factor, how the patient perceives her own capacity to control her behavior, leading to a new conceptual framework, the Theory of Planned Behavior (Ajzen 1985). According to this theory, the triggering of the behavior depends on the presence of particular circumstances or on the possession of resources (for example, time, money, a certain skill, cooperation of other people).

These theories have been applied to behavioral changes such as quitting smoking, beginning a physical activity, a diet, safe sex practices or adherence to a treatment for hypertension, bipolar disorder or urinary infection.

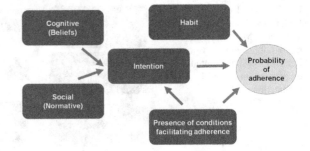

Fig. 2.3 Theory of Interpersonal Behavior and adherence. Modified from Triandis (1979) © Nebraska University Press

2.2.3 Theory of Interpersonal Behavior

In the *Theory of Interpersonal Behavior* (Fig. 2.3), developed in the late 1970s by Harry Triandis, three factors participate in the genesis of a behavior: The strength of habit in performing a certain behavior, the intention to perform it, and the presence of conditions that make performing the behavior easy or difficult.

This theory's major contribution is the importance accorded to the strength of habit: The degree of a behavior's automaticity (Triandis 1979). Later in this book, I will discuss in detail the crucial role that habit plays in patient adherence.

2.2.4 Leventhal's Self-Regulatory Model

Leventhal's theory (Leventhal et al. 1997) maintains that there is a regulatory cycle originating with the patient's representation of her illness, and proceeding to the measures that she takes. For the patient, it is a question of solving the problem posed by her illness or any other threat to her health. The patient responds in three stages: an interpretation of her illness, which can be triggered by internal signals (symptoms) or external signals (a doctor's diagnosis); the choice of adjustment measures or coping; and finally the evaluation of the results of her action—which, in turn, may modify her initial interpretation.

Adherence or nonadherence can be interpreted as one strategy among many of coping, each used to deal with the disease as perceived by the individual. For example, one may take an aspirin as a strategy to relieve a headache, and this strategy may be chosen thanks to the individual's belief that aspirin is usually a quick cure for headaches. If during the stage of evaluation the patient notices that the pain persists, she may change her strategy of coping (take a stronger pain medication) or reevaluate her representation of the illness (if the aspirin didn't work, maybe it's something more serious). According to this theory, the interpretation of the illness is based on a holistic picture that takes into account the problem's identity (what illness do I have?), its causes (how did this happen?), its consequences (what might happen?), and its curability (will this treatment work?).

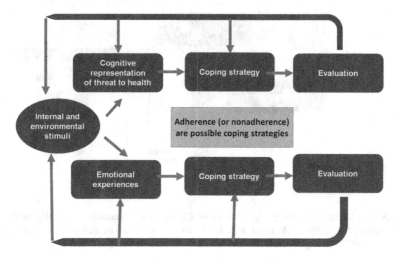

Fig. 2.4 Self-Regulatory Model and adherence. Originally published in Lange and Piette (2006). Modified with kind permission of © Springer 2006. All Rights Reserved

This theory, developed by Howard Leventhal in the 1970s, is unique for explicitly introducing two parallel paths for the three stages, one cognitive and one emotional (Fig. 2.4).

2.2.5 Transtheoretical Model of Change

The Transtheoretical Model of Change (Fig. 2.5) (Prochaska and DiClemente 1983; Prochaska and Norcross 1994) describes the different stages leading up to the adoption of a behavior. Developed by James Prochaska in the beginning of the 1980s, this model has been used to understand various behaviors: Smoking, alcoholism, drug addiction, routine exercise, weight loss, condom use, sun-screen use, mammography screening, and others. The model has also been called *transtheoretical* because it is a synthesis of the different psychological theories that were used at the time.

While other theories describe the adoption of a behavior as an event (stopping drinking, quitting smoking, beginning a diet), this model gives a progressive description and identifies five stages in the process leading up to the adoption of a behavior. Change is seen as the endpoint of an evolving process.

In the *precontemplation* stage, the individual is not conscious of having a problem, and so she has no intention to modify her behavior in the foreseeable future. A patient may be unaware of her problem thanks, for example, to a lack of information, or because she refuses to believe that there is a problem, or because she has already tried to resolve it, has failed and has given up. During this period, the individual avoids talking, thinking, and obtaining information about the problem,

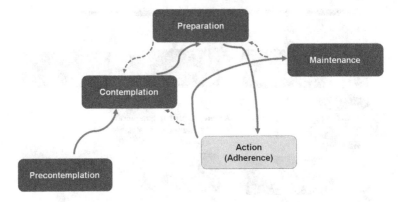

Fig. 2.5 Prochaska's Transtheoretical Model of Change and adherence. Originally published in Prochaska et al. (1992). Modified with kind permission of © The American Psychological Association 1992. All Rights Reserved

and is deaf to the words of others on the subject—it couldn't be otherwise, she knows nothing of it! In the pre-contemplation stage, the person either never heard about the problem, or, through a process of denial, refuses to see that there is a problem.

In the *contemplation* stage, the individual has begun to realize that there is a problem and considers doing something about it. For instance: She has weighed the benefits and drawbacks of taking action, but has not yet reached a decision; she puts off the decision to another day. Unfortunately, this state of procrastination can last a very long time.

During the *preparation* stage, the individual intends to act in the near future, and studies the ways of resolving the problem: She talks to her doctor, buys books on the subject, picks a start date for her diet or exercise plan, etc. The *action* stage is when the individual actually changes her behavior; and while the change in behavior may be quite dramatic—the alcoholic who puts down the bottle after years of steady drinking, for example—this model suggests that it is preceded by a long, and sometimes painful, germination.

Finally, the *maintenance* stage is the more or less prolonged period of time when an effort is required to avoid relapse.

Progression is seldom linear. There are frequent steps back, for instance with brief relapses. Usually the regression does not go all the way back to the *pre-contemplation* stage, but stops at the *preparation* or *contemplation* stages.

This model's primary contribution is to show that different interventions are needed at different stages of change. It is essential to establish where a patient may be on this journey before deciding how to intervene. The individual's position can be determined by questioning her about the arguments she is currently considering for and against the new behavior, or by evaluating where the individual places herself on the spectrum of self-efficacy, and in the perception of her vulnerability to temptation.

2.2.6 *The Reversal Theory*

Another theory, developed by Michael Apter in the beginning of the 1980s, the *Reversal Theory* (Apter 1982), maintains that an individual's perception of her situation can reversibly waver between two opposite states. For instance, a *'telic'* state is opposed to a *'paratelic'* state. The *'telic'* state (from Greek *telos*, goal) is a state of mind oriented towards the future, in the context of serious long-term projects. The 'paratelic' state, on the other hand, is one where the current activity is being enjoyed for itself, for the immediate pleasure it can bring. The other pairs of mental states are the conforming vs. negativistic states (we once again find the normative beliefs of the models described above), the mastery versus sympathy states (the world is seen either as a battle field or as a place open to generosity) and the autic versus alloic states: They refer to whether one is motivated by self interests (personal accountability and responsibility) or by the interests of others (altruism and transcendence).

2.3 Limitations of Psychological Models

The models just reviewed demonstrate statistical correlations between mental states and certain behaviors; but, as is well known, statistical correlations do not imply causal relations. In the foregoing illustrations, then, the arrows connecting mental-state-boxes to behavior-boxes represent only associations. On this basis, we can at most predict that if an individual, let's say Jane, believes that smoking is bad for her health, then Jane has a better chance of quitting smoking than if she does not hold this belief. But such a prediction remains simply statistical: It only indicates that belonging to the group of people who hold this belief gives Jane a better chance of belonging to the group of people who quit smoking than to the group of people who do not. And if one day Jane really does quit smoking, it does not follow that she did it because of this belief. She may have done it for a completely different reason, for example to please her daughter or because the price of cigarettes went up. Jane could also continue smoking, even though she believes it's bad for her health. She might be just as convinced that if she were to quit, she would gain thirty pounds like her neighbor, an idea that's unbearable to her. These models therefore have a major limitation: They cannot explain *why an individual* is or is not adherent to the medical advice that she is given. At the level of the individual, specific behaviors remain wholly unexplained and mysterious in these models.

To illustrate this problem, consider a taxi driver, Jeremy, who stops at a red light: The passenger in the back seat understands why they stopped, and he does not need to ask. But let us consider what a statistical study might tell us about this case. It would only show that *most people* stop when the light turns red. However, this does not mean that *John* stopped because the light was red—it might be that John never respects the law. No, that day John stopped because the bakery at the corner was open and he wanted to buy a pastry (Descombes 1995). Even though it is a fact that the great majority of people stop at red lights, if we are interested in *this particular event*, then this fact has no explanatory value. What explains why he stopped are *John's reasons*.

Can we perhaps find specific, non-statistical explanations for patient adherence? Can we formulate a theory according to which an individual's mental states *cause* her behaviors, just as insulin actually lowers blood glucose levels—not just as a statistical probability, but by virtue of a *mechanism*? Is it possible to describe the *mental mechanisms* of adherence to long-term therapies?

2.4 A New Perspective

If we seek a full explanation of adherence, we must change our point of view entirely. We must return to the individual; the resulting account will be unavoidably subjective (and maybe lose its "scientific" value), but it will allow us to infer conclusions applicable to particular individuals—nearly always the foremost concern of the practicing physician. In other words, behavior averaged out over a population is no longer sufficient for our investigation; rather, we shall focus on *what a given patient is actually doing and why (i.e. for what reasons) she is doing it*. We seek to establish a theory that defines what is meant by the 'reasons of care' and which shows how these reasons bring about caretaking—therapeutic—actions: A causal relationship between Mind and Care. Our theory must grasp the mechanisms of adherence at the level of the individual patient; only if we can achieve these aims will we have a genuine theory of care.

The models described earlier suggest a simplistic, stimulus-response behavioral schema—not coincidentally, the schema most amenable to quantitative psychological research. However, as effective as behaviorist models have been in explaining certain phenomena, they fall short of the mark with complex, real-world behaviors. In behaviorist models, human behavior is no different than, say, the solubility of sugar: When it is added to water its behavior is to melt. Reactions of these sorts, however, are not the same as *actions*—and it is actions which concern us, not just behaviors.

One feature which distinguishes a behavior from an action is the quality of intentionality. Just what do philosophers of mind mean by this concept of intentionality? Alfred Mele, a philosopher of mind, wrote:

> Remove the intentional altogether from intentional action, and you have mere behavior: brute bodily motion not unlike the movement of wind-swept sand on the shores of Lake Michigan (Mele and Moser 1994).

And Jean-Paul Sartre had noted that

> We should observe first that an action is on principle *intentional*. The careless smoker who has through negligence caused the explosion of a powder magazine has not *acted*. On the other hand the worker who is charged with dynamiting a quarry and who obeys the given orders has acted when he has produced the expected explosion; he knew what he was doing or, if you prefer, he intentionally realized a conscious project (Sartre 2003, 559).

There are two good reasons to examine adherence from the angle of action rather than behavior. First, it makes us consider each act of the treatment separately instead of confining ourselves to global patterns in behavior. This reflects reality: As we saw earlier, adherence is not an all or nothing phenomenon, it can sometimes have what we could call a regional character, as in the example of patients who take their medication but do not stay on a diet or quit smoking, even if we also noticed that these behaviors are often linked. Second, while the mechanisms underlying behavior are not self evident (we can see the sugar melting, but not its solubility), it is much easier to analyze the driving force behind an action; this analysis shall be the subject of this book.

In order to respect this distinction, we must from now on use a different, novel vocabulary, a *philosophical* vocabulary. Here, agency is treated as an event, independent of the investigator, which depends on an individual's intentional performance for certain 'reasons' that are her own. Among the 'reasons' there are, of course, 'mental states' such as knowledge, skills, beliefs, emotions, desires. In the philosophical vocabulary these mental states are called 'intentional', meaning that they have a 'content'. For instance, 'exercise makes one lose weight' and 'lose a few pounds' are, respectively, the contents of the belief and the desire in the thought: 'I believe that *exercise makes one lose weight* and I want to *lose a few pounds*' and this thought leads me to join a gym. This thought is the *reason* for this action. Having this definition of intentional mental states in mind, we may want to propose an intentionalist model of adherence and nonadherence to replace the behaviorist one, i.e. the "classic" view.

But the patient must not only perform the act of taking her pill; she will have to do it every day as long as it is necessary for an acute illness, and often for the rest of her life, in the case of a chronic disease. It is of little use to take the pill only once or only once in a while, just as driving under the influence of alcohol or failing to buckle one's seatbelt are not to be avoided only from time to time. The patient must acquire a behavior consisting of first accepting her treatment and then of accepting to perform it, if not each time, then at least of getting used to performing it as often as possible. This behavior is composed of *repeated actions*. To acquire such a behavior boils down to *usually* performing the acts (actions) of the treatment; thus, it becomes appropriate to invoke *habits* in the explanation of adherence. Following this analysis, we are tempted to replace the classic definition of adherence, "the concordance between the behavior of the patient and the medical prescriptions" by something like: "Accepting to repeatedly perform all the recommended health oriented actions". In the case of a chronic disease, it will be a long-term health goal.

2.5 In Search of Mental Mechanisms in Psychology and Philosophy

Attempting to describe adherence and nonadherence in terms of *repeated actions* rather than in terms of behavior leads us away from the domain of psychology, 'the science of behavior'. As Pascal Engel notes,

> We would search in vain among scientific psychology for a discipline that could be called 'psychology of action'. What everyday speech calls actions is redefined either by the psychology that considers only their corporal or physical aspect, such as behavioral

psychology, or by psychology in general, which considers actions from the angle of general traits such as personality, and that would be the psychology of 'conduct' (Engel Engel 1996, 146–147).

On the other hand, concepts developed by the philosophy of action are now available to our analysis. The essence of analytic philosophy is well described in one of Ramsey's last essays:

> Philosophy must be of some use and we must take it seriously; it must clear our thoughts and so our actions. Or else it is a disposition we have to check, and an inquiry that this is so (…) In philosophy we take the propositions we make in science and everyday life, and try to exhibit them in a logical system with primitive terms and definitions, etc. (…) In order to clarify my thought the proper method seems to be simply to think out with myself 'What do I mean by that?' 'What are the separate notions involved in this term?' 'Does this really follow from that?' etc., and to test identity of meaning of a proposed *definiens* and the *definiendum* by real and hypothetical examples. (…) We are driven to philosophize because we do not know clearly what we mean (Ramsey 1990, 1–6).

As we saw above, the different models describing the adoption of a health behavior have been developed in order to try to explain certain observations: For instance, the observation of a link between a particular health belief and quitting smoking. This link is represented by an arrow connecting two boxes, one representing the belief, the other representing the behavior—quitting smoking. But one could follow Ramsey's recommendation and ask: "What do I mean by belief?" "does quitting smoking really follow this belief?" *Is it correct to put an arrow between the two boxes*, in other words, are we justified in assuming that a certain belief *causes* one to quit smoking?

To answer these questions, are we not driven to philosophize?

We could formulate the question differently: How is it that one *never* finds *any* references to philosophers in the works on psychology of health or in the numerous books dealing with adherence to treatment? I believe this is in part due to the separation of disciplines and the lack of interdisciplinary work and also perhaps to historic reasons: As we saw, the major psychological models date to the 1950s–1970s. The works of analytic philosophy, which I will quote in the next part of this book, are generally more recent. Let's recall that the classic paper by Donald Davidson "How is weakness of will possible?" was first published in 1970, that the papers on weakness of will and *akrasia* by Gary Watson and Amelie Rorty were published in the late 1970s and that Alfred Mele published his book on *akrasia* in 1987. Similarly, the concepts of intertemporal choice and of multiple self, which are used to explain the weakness of will, derived from George Ainslie's first publications in the early 1970s and were mainly popularized in "The Multiple Self", edited by Jon Elster in 1986, and in his "Picoeconomics" published in 1992, the year when the first textbook on intertemporal choice, "Choice over Time" was edited by Jon Elster and George Loewenstein. Or, said in other words: The question on patient adherence I ask is essentially metaphysical—concerning the nature of an individual's actions, while the approach by psychologists is maybe more sociological.

2.6 Observation, Explanation and Mechanisms

So it can be *observed* that a particular belief is related to a refusal to quit smoking. Can we be content with *observing* this relationship without trying to understand its *mechanism*? Obviously not. If one wants to improve patient adherence, it is necessary to understand its mechanisms, exactly as discovering the mechanisms of diseases have made it possible to develop new therapies. The question is then, is it possible, when dealing with the mind, as in somatic psychology, to speak of *mechanisms*?

Before continuing, two precautions are in order. First, we will not describe but only allude to the neurophysiological mechanisms underlying psychological processes. This is the subject of neuroscience, even though, as we shall see later in the book, bridges are being built: We are beginning to be able to precisely locate the cerebral centers involved in the genesis of decisions or of emotions (Damasio 1994; LeDoux 1996), for instance. Our task will be limited to describing mental states *logically*: In brief, they are dispositions toward certain expressions, in word or in deed. For instance, Peter's fear that he is suffering from hypoglycemia may be expressed in an assertion—Peter says: "I'm afraid of hypoglycemia", or as an action—John reduces his insulin dose.

Second, in contrast to the natural sciences, we shall not be concerned with laws. The logical interactions of 'mental states', assertions, and actions don't have the form of laws, where A always brings about B. In the 'physiology of mind', A may bring about B *or* C, where C might be the opposite of B. For instance, the alcoholism of parents may lead to the children being alcoholics *or* sober, fear may lead to immobility *or* flight *or* fight. Thus, rather than use the word laws, it is better, as suggested by Jon Elster, to speak of mechanisms; these can be used afterwards to explain the observed behavior (Elster 1998, 45–73; Elster 2003, 25–82). Here is Elster:

> Are there lawlike generalizations in the social sciences? If not, are we thrown back on mere description and narrative? In my opinion, the answer to both questions is No. The main task of this essay is to explain and illustrate the idea of a *mechanism* as intermediate between laws and descriptions. Roughly speaking, mechanisms are frequently occurring and easily recognizable causal patterns that are triggered under generally unknown conditions or with indeterminate consequences. They allow us to explain but not to predict (Elster 1998).

2.7 Patient and Agent

Let's quote Descartes' *The Passions of the Soul*:

> To begin with, I take into consideration that whatever is done or happens afresh is generally called by the Philosophers a Passion with respect to the subject it happens to, and an Action with respect to what makes it happen. Thus, even though the agent and the patient

are often quiet different, the Action and the Passion are always a single thing, which has these two names in accordance with the two different subjects it may be referred to (Descartes 1989, 19).

In his book, *Le Complément de sujet*, Vincent Descombes comments:

> Descartes here evokes the general idea of an event: something that is done or happens. He notes that an event can be attributed to a patient, in which case it is called his 'passion' (in the physical sense of 'to suffer a change') or it can be attributed to an agent, in which case it is called an 'action' (Descombes 2004, 54–55).

In other words, a 'patient' is a subject to whom events happen, whereas an 'agent' is a subject who brings about that they happen.

Neither Descartes nor Descombes, it seems, are thinking of the most usual sense of the word 'patient', when used in the medical field. And yet putting these passages in our context suddenly gives them a somewhat strange resonance: Is the 'patient', in the medical sense of the word, condemned to remain a 'patient', in Descartes' sense of the word, one for whom the event of her illness (the new event) 'happens' and that she will then 'suffer'? How can the ill individual become the 'agent' performing 'actions', the *events* of her treatment?

This is the question this book asks.

References

Ain KB, Refetoff S, Fein HG, Weintraub BD. Pseudomalabsorption of levothyroxine. JAMA. 1991;266:2118–20.

Ajzen I. From intention to actions: a theory of planned behavior. In: Kuhl J, Beckmann J, editors. Action-control: from cognition to behaviour. Heidelberg: Springer; 1985. p. 11–39.

Anderson CJ. The psychology of doing nothing: Forms of decision avoidance result from reason and emotions. Psychol Bull. 2003;129:16–139.

Apter MJ. The experience of motivation: the theory of psychological reversals. London: Academic Press; 1982.

Bader A, Kremer H, Erlich-Trungenberger I, Rojas R, Lohmann M, Deobald O, Lochmann R, Altmeyer P, Brockmeyer N. An adherence typology: coping, quality of life, and physical symptoms of people living with HIV/AIDS and their adherence to antiretroviral treatment. Med Sci Monit. 2006;12:CR493–500.

Baumeister RF, Bratslavsky E, Finkenauer C, Vohs KD. Bad is stronger than good. Rev Gen Psychol. 2001;5:323–70.

Becker MH, Maiman LA. Sociobehavioral determinants of compliance with health and medical care recommendations. Med Care. 1975;13:10–24.

Brehm JW. A theory of psychological reactance. New York: Academic Press; 1966.

Briesacher BA, Andrade SE, Fouayzi H, Chan KA. Comparison of drug adherence rates among patients with seven different medical conditions. Pharmacotherapy. 2008;28:437–43.

Daley DC, Zuckoff A. Improving treatment compliance, counseling and systems strategies for substance abuse and dual disorders. Center City: Hazelden; 1999.

Damasio AR. Descartes' error. Emotions, reason and the human brain. New York: G. P. Putnam's Sons; 1994.

Descartes R. The passions of the soul. Indianapolis: Hackette Publishing Company; 1989.

Descombes V. La Denrée mentale, Éditions de Minuit; 1995. p. 59–62.

Descombes V. Le Complément de sujet, enquête sur le fait d'agir de soi-même, nrf essais, Gallimard; 2004.

Elster J. A plea for mechanisms (Chap. 3). In: Hedstrøm P, Swedberg R, editors. Social mechanisms: an analytical approach to social theory. Cambridge: Cambridge University Press; 1998. p. 45–73.

Elster J, Skog O-J, editors. Getting hooked, rationality and addictions. Cambridge: Cambridge University Press; 1999.

Elster J. Un plaidoyer pour les mécanismes. In: Proverbes, maximes, émotions, P.U.F.; 2003. p. 25–82.

Engel P. Philosophie et Psychologie. Gallimard; 1996.

Fishbein M, Ajzen I. Belief, attitude, intention, behaviour. Don Mills: Addison-Wesley; 1975.

Fogarty JS. Reactance theory and patient noncompliance. Soc Sci Med. 1997;45:1277–88.

Fogarty JS, Youngs GA Jr. Psychological reactance as a factor in patient noncompliance with medication taking: a field experiment. J Appl Soc Psychol. 2000;30:2365–91.

Gilovich T, Griffin D. Introduction heuristics and biases: then and now. In: Gilovich T, Griffin D, Kahneman D, editors. Heuristics and biases, the psychology of intuitive judgment. Cambridge: Cambridge University Press; 2002. p. 1–18.

Harrold LR, Andrade SE, Briesacher BA, Raebel MA, Fouayzi H, Yood RA, Ockene IS. Adherence with urate-lowering therapies for the treatment of gout. Arthritis Res Ther. 2009;11:R46.

Janz NK, Becker MH. The health belief model: a decade later. Health Educ Quaterly. 1984;11:1–47.

Kahneman D, Tversky A, editors. Choices, values and frames. Cambridge: Cambridge University Press; 2000. p. 1–16.

Lange LJ, Piette JD. Personal models for diabetes in context and patients' health status. J Behav Med. 2006;29:239–53.

Laplantine F. Anthropologie de la maladie. Bibliothèque Scientifique Payot; 1997.

LeDoux J. The emotional brain. New York: Touchstone; 1996.

Leventhal H, Falconer Lambert J, Dieffenbach M, Leventhal EA. From compliance to social-self-regulation: models of the compliance process. In: Blackwell B, editor. Treatment compliance and the therapeutic alliance. Amsterdam: Harwood Academic Publishers; 1997. p. 1997.

Meichenbaum D, Turk DC. Factors affecting adherence. In: Facilitating treatment adherence. New York: Plenum Press; 1987.

Mele AR, Moser PK. Intentional actions. Noûs. 1994;28:39–68.

Prochaska J, DiClemente C. Stages and processes of self-change in smoking: toward an integrative model of change. J Consult Clin Psychol. 1983;5:390–5.

Prochaska JO, DiClemente CC, Norcrosss JC. In search of how people change. Applications to addictive behaviors. Am Psychol. 1992;47:1102–14.

Prochaska J, Norcross JC. Person-centered therapy. In: Systems of psychotherapy, a transtheoretical analysis. Pacific Grove (CA): Brooks/Cole; 1994.

Ramsey F. In: Mellor DH, editor. Philosophical papers. Cambridge: Cambridge University Press; 1990.

Reach G. Linguistic barriers in diabetes care. Diabetologia. 2009;52:1461–3.

Reach G. Obedience and motivation as mechanisms for adherence to medication. A study in obese type 2 diabetic patients. Patient Prefer Adherence. 2011a;5:523–31.

Reach G. Treatment adherence in patients with gout. Joint Bone Spine. 2011b;78:456–9.

Reach G. A psychophysical account of patient nonadherence to medical prescriptions. The case of insulin dose adjustment. Diabetes Metab. 2013;39:50–5.

Sartre JP. Being and nothingness, an essay on phenomenological ontology (trans. Barnes HE). 2nd ed. London: Routledge; 2003.

Triandis HC. Values, attitudes and interpersonal behavior. In: Page MM, editor. Nebraska symposium on motivation, vol 27; 1979. p. 95–259.

Tversky A, Kahneman D. Judgment under uncertainty: heuristics and biases. Science. 1974;185:1124–31.

Chapter 3
Intentionality

Abstract This chapter presents a taxonomy of the sorts of mental events that explain the why behind actions—paradigmatically, the agent's beliefs and desires. First, there are propositional attitudes: These sorts of mental states are formulated in terms of an agent's attitude toward a proposition. Such a mental state that has content is also called an intentional state: For instance, emotions are intentional state, and this differentiates them from sensations such as pain and from moods such as being sad, brooding or cheerful. In this chapter, I describe mind as a jigsaw puzzle— a puzzle without borders which can grow indefinitely. Each new belief must find a place where it fits with adjacent beliefs and also coheres with the emerging picture. The second part of this chapter is aimed to describe the place of these mental states in agency. A desire-belief pair allows rationalizing the action: This is the reason I could give if someone asked me—Why did you do this? Donald Davidson, in his Causal Theory of Action, went one dramatic step further: He contended that this reason is also the real cause of the action. In other words, the reason of an action causes this action, like insulin causes a decrease in blood glucose. The realization of an action is connected to a set of pertinent mental states, the person acting after having all well considered. This conception of action eschews deterministic laws, but does acknowledge the force of mechanisms, in a quasi-physiology of Mind.

In what follows, it will be convenient to have a fairly systematic taxonomy of the sorts of mental events that explain the *why* behind actions—paradigmatically, the agent's beliefs and desires. Although our knowledge of the motor system is impressive, we do not yet have a firm understanding of the neurophysiological processes underpinning mental events. While waiting for a neuro-physiological description to become possible, we can only stick to the humility of Leibniz (1698), who wrote about the impossibility of understanding what perception is:

> Moreover, it must be confessed that perception and that which depends upon it are inexplicable on mechanical grounds, that is to say, by means of figures and motions. And supposing there were a machine, so constructed as to think, feel, and have perception, it might be conceived as increased in size, while keeping the same proportions, so that one might go into it as into a mill. That being so, we should, on examining its interior, find only parts which work one upon another, and never anything by which to explain a perception (Monadology 17).

© Springer International Publishing Switzerland 2015

G. Reach, *The Mental Mechanisms of Patient Adherence to Long-Term Therapies*, Philosophy and Medicine 118, DOI 10.1007/978-3-319-12265-6_3

Nonetheless, over the past century or so, philosophers of mind have articulated the *logical* relations of the mental mechanics behind actions. This analysis is tentative, of course, but its conceptual apparatus is rich, and it permits us to develop an account of adherence and nonadherence that is precise enough for now (insofar as the concepts from philosophy of mind are precise enough) and which may yield into deeper, more concrete explanations as philosophy of mind, cognitive psychology, and neurophysiology reach mutually reinforcing conclusions.

The aim of this chapter is to present a tentative description of mental states (beliefs, desires, emotions and the like), showing how they interact not only to represent an explanation of our actions, but really *to cause them*.

Medical doctors, who are familiar with biochemical and physiological events, but not so with philosophical concepts, may discover in this "philosophy of mind" a sort of "*physiology* of mind"; and they may find here an analogy with old friends like substrates and enzymes, hormones and receptors, etc. They will discover that it is possible to construct a "physiology of mind" in which mental events are causally linked, exactly in the same way that insulin is the cause of a decrease in blood glucose. Next they will understand that one can discuss nonadherence to long-term therapies as if it were an abnormality in this "physiology", just like diabetes is an abnormality in the regulation of blood glucose. On the other hand, philosophers who are convinced of this theory of mind will find here support for their conviction: Their theories have a fruitful application.

3.1 What Is 'In Your Head'

What types of mental states are there? First, there are what Bertrand Russell called propositional attitudes (On the genesis of the notion of propositional attitudes, see Maslin 2001, 16–17). These sorts of mental states are formulated in terms of an agent's *attitude* toward a *proposition*. Consider Gerard's belief that aspirin will make him sick. The proposition, also called the *content*, follows the 'that'; in this case, '*aspirin will make [Gerard] sick*'. And Gerard's attitude toward this proposition is characteristic of *belief*. These are the logical components of a propositional attitude. The content of the propositional attitude can eventually be expressed (in an assertion) or evaluated (by judgment). One can have various attitudes toward the same content: This is characteristic of mental states, and it differentiates them from objects. A proposition (for example, "aspirin can reduce fever") can be believed, desired, feared, regretted. Suppose that as a decidedly tepid form of revenge, you've sought to give Harry a fever, and you hope it shall last the few hours of your visit. As you see him take two aspirin with dinner, you may *fear* what he *desires*, namely, that *aspirin can reduce fever*. And, later, when Harry's fever is reduced, you may then *regret* what you previously feared—again, that aspirin can reduce fever. Throughout your changing attitudes about it, though, aspirin remains chemically the same.

A mental state that has content is also called an *intentional* state. Here this term must not be taken in the usual sense of intention. It simply means that the state is "directed at something", which is its content. For instance, belief and desire are archetypes of intentional mental states because one cannot simply *believe* or simply *desire*. One believes or desires *something*: I believe *it is raining*, I desire *this book to be read*.

Belief and desire are both intentional states because these attitudes have a content; still, there is an important difference between the two. When I believe something, for instance that it's raining, the content of my belief may be true or false. If it is raining, I believe something true; if not, then my belief is false. Clearly, it depends on whether the world is as the content of my belief represents it to be: Right now, it is either raining or it is not. If my belief is false, the error lies with my belief and not with the world; upon realizing my mistake, it is appropriate for me to change the content of my belief, adjusting my mind to the state of the world. Intuitively, beliefs *are supposed to be true*; and any that aren't ought to be discarded. Ideally, one's beliefs "fit" the way the world is. Such observations have inspired John Searle's claim that beliefs have a "mind-to-world direction of fit" (Searle 1983, 8).

On the other hand, if not a single person has read this book, should I change my desire to "I hope no one ever reads this book"? Of course, not. Rather, it is appropriate for me to *change the world* so that it satisfies my desire. Here, it is the state of the world that is responsible for satisfying or not satisfying a desire's content and thus for fulfilling or not fulfilling my desire. I could, of course, *renounce* my desire, but it is not the same thing as accepting the fact that it is not fulfilled. Deep down, the goal of my desire is to change the state of the world. According to Searle, desire, as opposed to belief, always has a "world-to-mind direction of fit". I can reach my goal by giving you my book to read. Once you have read it, I will have adjusted the world to fit my state of mind: There will be someone in the world who has read it.

Searle illustrates this dynamic interplay between the mind and the world with a vivid example showing how an adjustment, in principal a symmetrical concept, can have two possible directions. When Cinderella is shopping for new shoes before the ball, the direction of fit is from slipper to Cinderella. If the shoe doesn't fit Cinderella's foot, it's the wrong shoe. When Prince Charming is looking for the girl who lost the slipper that evening on the steps of his palace, however, the direction of fit is from Cinderella's foot to the lost glass slipper. If the shoe doesn't fit Cindarella's foot, then her's is the wrong foot.

3.1.1 The Different Types of Intentional Mental States

There are many more mental states besides desires and beliefs. These two are akin to the primary colors, though: In various combinations, they make up many other mental states, such as knowledge, surprise, fear, hope, regret, pride, shame, and maybe still other attitudes. 'In my head' there are, besides beliefs and desires, things that I know, memories, people, objects or events that I am observing at this moment and that I am aware of, sensations such as pain, and finally, emotions.

Insofar as it is plausible that emotions are directed at an object, then, it is plausible that they are intentional states. This differentiates them from sensations such as pain and from moods such as being sad, brooding or cheerful (Clore and Gasper 2000, 12). Emotions are states such as love, hate, anger, sadness, pity, shame, disgust, etc., which often have a person as their object (Elster 1999b, 271) (I hate this guy) rather than a proposition (I hate that my country could be invaded). As shown by Jon Elster, emotions are defined not only by visceral excitement but also by the fact that they are influenced by cognitive factors (Elster 1999b, 328–331; Elster 2000, 135–191): Generally, they have an intentional object (I am jealous of *someone*), they can be ignited by a belief (I am jealous of him because I believe his professional position is superior to mine), they can provoke the appearance of another emotion (I am ashamed of being jealous); they can finally mutate into another emotion (rather than being jealous, which makes me ashamed, I prefer to be indignant at the thought that he could have attained this high position which, in my opinion, he does not deserve).

All these mental states are different from visceral sensations such as hunger, thirst, the need to urinate or to sleep, pain, exhaustion, nausea, anxiety, boredom, annoyance, etc., which are perceived as such, without always having a content: I can say 'where' I feel pain (my head), I can describe the sensation (it itches or tickles) but I cannot say I feel pain '*that*': While I can say—I think *that* (I am happy with the treatment I received from the doctor)—my thought has a content which starts with the bracket, this cannot be the case for a pain.

It is important to distinguish between the mental states that have content and those that do not. A desire to eat a piece of cake has content; feeling hungry does not. According to Searle's theory of Intentionality[1], mental states which have content can be verified by outside observers: If I believe it is raining, one can look outside and see; if I want to eat a piece of cake, one can watch to see if I eat the cake put in front of me. Content-free mental states cannot be verified by others: Only I know if I truly feel pain.

In practice, the distinction between having or not having content may be difficult to make. For example, I cannot verify if another person has pain, but I can certainly observe his demeanor, observe if he reaches for the aspirin, etc. Similarly, a person who claims to want to eat cake might not eat the cake I give him because he's allergic to that type of cake, or because he's on a diet, etc.

Elster argues similarly that

> these various motivational factors can be uncontroversially located on a continuum. At one extreme we have the noncognitive or purely visceral states of pain, drowsiness, etc. Next are the states that have intentional objects but are not otherwise shaped by cognition, such as hunger, thirst, and sexual desire. Further, there are cravings that have intentional objects and that can also involve cognitions in other ways. Then there are emotions, which often involve cognition in all three ways. At the other extreme of the continuum, there are motivational states that do not imply any arousal or viscerality at all, as in my calm decision to take an umbrella because I believe it will rain and I don't want to get wet (Elster 2000, 3).

---

[1] Conventionally, Searle writes the words Intentionality, Intentional, etc., with a capital I.

3.1.2 The Place of Pleasure

For Patrick Pharo,

> pleasure is at once a sensation, an emotion and a feeling. It is a sensation because it is usually associated with sensory data emanating from the five senses or possibly from senses we are unconscious of, like the vomeronasal organ sensitive to pheromones. It is an emotion because of particular bodily reactions expressed, for instance, by physical signs such as variations in cardiac rhythms or the temperature of the skin and certain neuroendocrine mechanisms. Finally, it is a feeling because it characterizes a certain quality of a past experience felt as pleasant (Pharo 2006, 224).

Although it has a mixed character, pleasure is closer to a non-intentional mental state; for it does not have a content, like usual intentional mental states. In this it is similar to pain. Pain is also a sensation whose neurobiological mechanisms are well known; it has an emotional component because it involves a visceral arousal, and it can be influenced by cognitive factors, even though it does not have a 'content'—we saw that I can say where I feel pain, but not that I feel pain 'that'. Similarly, I can say I feel pleasure and describe it, but cannot say I feel pleasure 'that'. To prove the non-intentional nature of pleasure, we can note its similarity to other mental states we have described: Pharo writes that pleasure is

> often associated with different concepts of feeling such as satisfaction, contentment, comfort, ease, well-being, relief, charm, enjoyment, delight, enchantment, gaiety, joy, jubilation, exultation, enjoyment, rapture, happiness, euphoria, felicity (…) At the end, the best way to avoid too limiting definitions of pleasure and to take into account its diverse manifestation would be to define it as an experience that one does not get tired of or only gets tired of when it no longer gives pleasure, i.e., when it is no longer a pleasure. We could also say that when the pleasure concerns beings who have the power of speech, it is an experience that always makes us say 'Again!' which explains why it is so difficult to give up something that gives pleasure and that satiation is but pleasure (Pharo 2006, 228).

3.1.3 What Mental States Do

Elster uses the term 'motivational factor': It expresses the fact that propositional attitudes tend to manifest in the form of an action or an assertion (the expression of their content); it is not always the case, though, just as sugar is soluble without actually having to dissolve to prove it (Engel 1995, 25).

The relationship between propositional attitudes and their tendency to assert their content or to drive action is described by the dispositional-functionalist conception of the mind. According to this conception of the mind, particular propositional attitudes serve as a transition between incoming information (for example, perceptions) and observable results (for example, actions or verbal expression) or other mental states. To give a simplified example, the belief that smoking is harmful and addictive may serve as the transition between the perceptions of a friend smoking and the action of asking her to quit. Again, let us stress that this is a logical description of mental processes and not a hypothesis of their neuro-physio-logical function (Maslin 2001; Heil 1998).

We can evoke the classic definition of a belief by Hume (1740)

> But its true and proper name is belief, which is a term that every one sufficiently under-stands in common life. And in philosophy we can go no farther, than assert, that it is something felt by the mind, which distinguishes the ideas of the judgment from the fic-tions of the imagination. It gives them more force and influence; makes them appear of greater importance; infixes them in the mind; and renders them the governing principles of all our actions (Hume, *A Treatise of Human Nature*, 1.3.7).

We can go even further. Ramsey argued that mental states have a quantitative dimension in addition to their more obvious qualitative aspects. Ramsey under-stood mental states as components of a map of our behavioral world, a "map of neighboring space by which we steer" (Ramsey 1990, 146). He also described it quantitatively—just as a map has qualitative (landscape features) and quantitative aspect (distances and geodesics): (1) "The degree of a belief is a causal property of it, which we can express vaguely as the extent to which we are prepared to act on it" (the more I am convinced that yellow mushrooms are edible, the more I will be prone to eat some), (Ramsey 1990, 64) and (2) "true beliefs are those that lead to the success of our actions, whatever the desire in question" (the more it is true that yellow mushrooms are edible, the less is the risk to eat them) (Dokic and Engel 2001, 71). Ramsey contended that the interrelationships between the degree of a belief, the truth of its content, and the success of the action to which it tends to lead were mathematically quantifiable.

Although the interrelationships between belief, content and action are (at least arguably) quantifiable, this by no means establishes a determinate causal relation-ship between them. It may be that my belief that yellow mushrooms are edible only sometimes is the cause of my eating one. Nevertheless, we can begin to see the outlines of a concept of a "physiology of mind"—albeit more complex and employing different mechanisms.

3.1.4 Holistic Conception of the Mind

Just as we cannot understand physiology without knowing how various physio-logical processes work together, so we cannot imagine mental states in isolation: According to Wittgenstein (1969),

> when we first begin to *believe* anything, what we believe is not a single proposition, it is a whole system of propositions. (Light dawns gradually over the whole). It is not single axioms that strike me as obvious, it is a system in which consequences and premises give one another *mutual* support (Wittgenstein *On Certainty*, 141, 142).

Mental states are essentially holistic, i.e., they form networks of knowledge, of beliefs, desires, hopes, regrets—in short the luggage consisting of all the contents of propositional attitudes, of which some are certainly innate, while others are collected during the voyage of life. It is the *all* that the amnesiac, like the Jean

Anouilh's 'traveler without luggage' coming back to his family, has lost. Let's say that his intentional states lost their contents.

This '*all*', that characterizes holism in the proper, etymological sense, must be understood, according to Vincent Descombes, as the rejection

> of atomism, meaning the idea that we will be able to reconstruct people's mental life by combining, through the association of ideas or the linking of signifiers, psychic atoms (Descombes 1996, 89, 97–103).

Because the '*all*', in as much as there is an 'all', contains more than the sum of its parts. Here we indeed find ourselves at the extreme opposite from the classic behaviorist conception of the black box, where there are no actions, just behavior which is supposed to be a response to a stimulus. Donald Davidson in his *Paradoxes of Irrationality* similarly describes the nature of mental holism:

> The meaning of a sentence, the content of a belief or desire, is not an item that can be attached to it in isolation from its fellows. We cannot intelligibly attribute the thought that a piece of ice is melting to someone who does not have many true beliefs about the nature of ice, its physical properties connected with water, cold, solidity, and so forth. The one attribution rests on the supposition of many more – endlessly more (Davidson 2004, 183).

3.1.5 The Background

The mental states that we have just described, intentional or not, conscious or not, still do not exhaust 'what is in your head'. John Searle asked what it would take to describe everything necessary for an act such as getting a cold bottle of beer from the refrigerator in order to drink it. After the beliefs and desires leading to relatively specific actions (the refrigerator door that must be opened, the thirst that one desires to quench, the glass taken out of the cupboard, the bottle opener, etc...), one would quickly come to the entities that are no longer intentional states, but rather capacities such as the capacity to recognize the door and open it, the fact that the table is solid enough to hold a bottle of beer. In sum, these are notions that are so obvious that we do not need to think about them; Searle groups these together under the term 'Background':

> The Background is rather the set of practices, skills, habits and stances that enable Intentional contents to work in the various ways that they do, and it is in that sense that the Background functions causally by providing a set of enabling conditions for the operation of Intentional states (Searle 1983, 141–159).

The networks of propositional attitudes are inserted in the framework of this Background. So Anouilh's Traveler without luggage can lose the content of his propositional attitudes and his intentional states, to use Searle's terminology. But the traveler keeps his Background, as is proven simply by his coming on to the stage: He is capable of walking and talking, and thanks to the continuity of the Background that acts as a matrix, he will quickly be able to give new propositional content to different attitudes when forging new social relationships with what he believes to be his new family.

3.2 A Mental Puzzle and Its Formation

The formation of a new durable belief (for instance, 'I believe that I am ill') is the result of several assessments: Of probability, taking into account everything I know; of credibility, based on the evaluation of the reliability of the sources at my disposal; and finally, of plausibility, examining whether I can explain the phenomenon that is the object of the belief (Fridja and Mesquita 2000, 69). In order for a given belief to be able to durably integrate itself into our mind it must be reasonably consonant with all the propositional attitudes that holistically compose our mind to find an acceptable place there. The content of the belief not only has to be accepted, it must also not be rejected. As Karl Popper notes when speaking of scientific knowledge, it is particularly important, before accepting a new theory, to confront the arguments against it, rather than being satisfied with the arguments in its favor. For Robert Nozick (Nozick 1993, 72–73),

> Rationality involves being responsive to relevant factors, to all and only the relevant factors. It is an additional thesis that the relevant factors are *reasons*. It is a still further thesis – one that may not hold in all domains – that these reasons divide neatly into the two categories of *for* and *against*.

Here Nozick uses the definition of the concept of reason given by Bertrand Russell: "It signifies the choice of the right means to an end that you wish to achieve. It has nothing whatever to do with the choice of ends (Russell 1954)."

Coming to accept a new belief is like fitting a new piece in a jigsaw puzzle—but a puzzle without borders which can grow indefinitely. Each new belief must find a place where it fits with "adjacent" beliefs and also coheres with the emerging picture. It is in this way that by successive additions of beliefs the holism of our mind is enriched and creates possible spaces for the potential integration of new beliefs.

Unlike a growing jigsaw puzzle, however, the accretion of beliefs happens passively—or at least without voluntary control. We cannot simply decide to hold a new belief, though we may act in ways to foster new beliefs, such as reading books by a certain author, or cultivating an attitude of open-mindedness. As much as we feel we "own" our beliefs, it seems that what we really control are some of the conditions for belief accretion.

On the other hand, we can voluntarily *refuse* to believe something, after having gone through some degree of assessment. For example, I might reject the idea that a popular new diet will help me lose weight when I follow the diet and I lose no weight. Or, I might briskly reject a belief on the basis of visceral repulsion ("I refuse to believe she would have done that!") But we cannot do the same for the *formation* of a belief: One cannot, as Pascal Engel puts it, suddenly *decide* to believe something, "like deciding to go out of town for the weekend", although whether one can want to believe remains a disputed question (Losonsky and Dupuy 2000, 101–143).

The image of a puzzle metaphorically describes this asymmetry: The assessments of probability, creditability and plausibility mentioned earlier function to refuse or accept a new belief, taking into account the empty spaces in the puzzle, but chiefly the form of the outline of the existing pieces (by 'form of the outline' I mean not only their special form, but also the strength of attraction or repulsion—according to the idea that there are arguments for or against—similar to a magnet). These assessments are made *according to the current state of the puzzle*; one cannot have a durable belief that is not attached to other pieces of the puzzle, floating in isolation. So it is the puzzle itself, in its current state, and not the will of the agent, that conditions the integration of a new belief: It will be refused, whether the agent wants it or not, if it does not find another piece to attach itself to. In short, one cannot force a piece to fit into the puzzle, but one can indeed establish that a piece has nowhere to fit (yet). This might be the meaning of the expression we sometimes use when speaking of something we cannot believe: 'It does not *fit*'.

The passive character of belief acquisition explains how the mind can have a certain degree of coherence: In other words, the fact that we display, at least generally, a certain degree of rationality (Engel 2000, 3). If it were possible for us to want to believe, we could voluntarily introduce into the puzzle of our mind a belief that would not have its place there. Davidson insists on the necessary character of this relative rationality; it is necessary for interpersonal communication:

> And among the beliefs we suppose a man to have, many must be true (in our view) if any are to be understood by us. The clarity and cogency of our attributions of attitude, motive, and belief are proportionate, to the extent of which we find others consistent and correct. We often, and justifiably, find others irrational and wrong; but such judgments are most firmly based when there is the most agreement. We understand someone best when we hold him to be rational and sage, and this understanding is what gives our disputes with him a keen edge (Davidson 2004, 184).

3.2.1 The Necessary Incompleteness of the Mental Puzzle

However, the puzzle of our mind cannot be complete: Some pieces are necessarily missing, although some of them might be deemed essential by an outside observer. The emotions in particular intervene by selecting among all the available information and emphasizing some of it (De Souza 1987, 195–198, Bower and Forgas 2000, 87–168), helping to avoid the pressure to choose from an overflow of information that would be paralyzing. Emotions seem to focus our "mental attention" on certain information. Neurophysiological evidence supports this claim. Patients suffering from lesions to the prefrontal ventromedian zone of the cerebral cortex exhibit both a decrease in the capacity to feel emotions and to make decisions.

Pierre Livet assigns to emotions a similar role in the *revision* of mental states: Emotions arise from an observation of the difference between what we thought concerning the state of the world and what we observe (for instance, I thought I was perfectly safe in the forest when the sight of a snake leads to the emergence of fear), and they lead to a revision of what we thought:

> To revise is to change the premises and the inferences that lead us to a conclusion proven to be false by newly learned facts (…). We propose to admit that when there is to be a

revision, the emotion will tend to reappear as long as the revision has not been completed. On the other hand, if we perform the revision and change our expectations, the emotions should disappear (Livet 2002, 23, 29).

As Livet points out, the existence of emotions may even be *necessary* for such a revision. Recall that, according to Searle, beliefs should be adjusted to fit the world while the world should be adjusted to fit one's desires. For Livet, emotions give us either the motivation to change the world or the motivation to change our previous desires. The emotions are not a sufficient condition of these changes, but they are a necessary condition. They adjust us to the world, like beliefs, but according to the adjustments that we desire on the part of the world. When we cannot change the world, the motivation to change our preferences wins, unless it turns out that it is just as difficult to change our preferences (Livet 2002, 77–79).

George Loewenstein et al. noticed that in traditional models of decision, primarily "consequentialist" (individuals make up their mind by appreciating the consequences of their choice), emotions would be essentially regarded as epiphenomena (Loewenstein et al. 2001). They proposed an alternative model where feelings play a causal role in behavior under risk conditions. This model may be pertinent in the context of patient adherence, as a number of actions performed in the framework of a treatment may, in the patient's mind, present a risk.

Loewenstein et al. introduced a distinction between anticipatory and anticipated emotions. The first are immediate, visceral, affects for example the fear, the anxiety or the dread, which one feels when one has to achieve an action which presents a risk. The anticipated emotions are not felt immediately, but are those which one imagines that one is likely to feel as a consequence of the decision. For instance, the importance of anticipated regret in decision can be demonstrated by the following empirical study: Imagine a $100 ski pass and three groups of participants. Participants of Group 1 missed a hypothetical opportunity to buy the ski pass for $40. Those of Group 2 missed an $80 price. Group 3 had no initial opportunity. All participants then had the opportunity to buy a pass for $90. Participants of Group 1 rated themselves as least likely to purchase the ticket. These results can be explained by regret avoidance (regret would be more important in Group 1) (Tykocinski et al. 1995).

In this model, individuals evaluate the possible outcomes of a choice at risk in a cognitive way, as in the traditional model, by taking into account the desirability and the probability of the outcome; anticipated emotions are part of the outcome. This cognitive evaluation has consequences, including emotions which interact with the results of the cognitive evaluation. Moreover, there are anticipatory "feelings" which can also be triggered in a fast, non-cognitive, way, according to the context. This "risk-as-feelings" model proposed by Loewenstein et al., represents therefore a description of rationality having two distinct pathways which can possibly be contradictory.

Desires also have a role to play in the way we collect information and form our beliefs (Elster 1999a, 16–17): We remember and believe more easily what we desire. David Pears gives the example of a smoker who easily convinces herself that smoking is not dangerous because she wants to continue smoking (Pears

1985, 62). We shall see that this role of desire can lead to irrational behavior. But even here it can also have a positive effect when it makes us persevere in a useful task by sweeping aside the arguments that could deter us.

This positive role of emotions and desires in the formation of beliefs has one important consequence: Emotions and desires, by acting as a filter, imply a necessary incompleteness of the holism of our mind. This mechanism explains why different individuals faced with the same situation will transform the stimuli according to their own emotions and desires, i.e., the ones that are already part of their mental puzzle, creating different compositions of new beliefs. One could imagine that the mental puzzle is like a collage, with each fragment being the product of unique, individual dynamics: Emotions, desires, existing beliefs, genetics, and so forth.

We have seen that there are numerous types of interrelations between mental states: The interrelations between beliefs and desires on the one hand, and the emotions (fears, hopes, regrets, pride, shame) on the other. We may add a third type of interaction: The interaction between current action and those which preceded it. Previous action does not determine current action, of course, but neither is it unrelated (Elster 1999a, 16–17). This last type of interaction is also noted by Nozick: "Beliefs about the world feed forward into actions, and the (perceived) results of these actions, along with other perceived facts, feed back, positively or negatively, upon the beliefs" (Nozick 1993, 99). Repeated prior actions etch a channel, as it were, in the mind's terrain.

Roy Baumeister et al. proposed a model in which emotions provide the basis for such a feedback (Baumeister et al. 2007). Suppose that a given action, following the application of an "if-then rule" (if I find myself in such situation, then I will do such and such) leads to an unpleasant outcome, inducing the occurrence of a strong negative emotion. The next time the rule will have to be applied, a short recollection of this feeling will occur like a flash, helping the individual to avoid the same mistake. It thus appears in this model that the role of the emotions, as opposed to what one believes intuitively, is not to cause the behavior, but to shape the cognitive process. Baumeister et al. noticed in an elegant way that one can understand, within the framework of this model, why one cannot control emotions: You cannot control them, *because the role of the emotions is precisely to control you.* This explains the "near-miss effect": There is more emotion after just missing your train by a few minutes than after missing it by half an hour. It is important that you remember this story vividly: If you miss your train by 3 min, then you may profitably regret dawdling over your second cup of coffee; n*ext time* you will skip this second cup and make the train. The emotions represent therefore a system of feedback the goal of which is to provide training and control of behaviors.

In conclusion, we can propose that the puzzle of our mind is formed by at least three mechanisms: (1) A completely passive mechanism that allows the rooting of new mental states in the vacant spaces between the previously accumulated pieces, and which adjusts our mind to the world around it. This first mechanism requires some congruence between the new piece and the ones already present; so there is a first screening that allows us to reject a new belief through a more or less conscious analysis of its plausibility, its probability and its credibility. (2) The second mechanism consists of refusing, according to the emotions and desires proper to the agent, certain pieces that normally could have been accepted or, on the contrary, assigning a special importance to some piece. We have already noted the benefits of such a process: It prevents the clogging of the mind that would inhibit it when making decisions (positive effect of emotions), and in some cases it could prevent us from giving up projects that will be beneficial in the long run (positive effect of desires). (3) Finally, there is the third mechanism, where the results of our actions may also lead to the modification of the mental puzzle. Here we come back to the notion of regulation present in the psychological model of adopting a health behavior proposed by Leventhal, described in the previous chapter, and similar to the mechanisms of feedback observed in physiology.

This is how we can represent the way our more or less filtered image of the world is formed. And rather than saying that an agent, by acquiring a new belief, adjusts her mind to the world, it would perhaps be more precise to say that this adjustment takes place at the level of her image of the world. The puzzle of the mind is not only a filtered image of the world around us. It is also the interface where some of its own components, the emotions and desires, and also the evaluation of the results of our actions, generate the remodeling that both allows and limits its continual enrichment. We will soon see that the very same mechanisms open doors for irrational phenomena, just as in medicine the same mechanisms are responsible for both physiological and pathological states.

We are now ready to ask the question: *Why do we do or not do something*? This will bring us back to the issue of patient adherence and non adherence.

3.3 Actions

Let us then propose that taking a pill is an action, and proceed to analyze the phenomenon of adherence (I take the pill as it was prescribed) or of nonadherence (I do not take the pill) from this angle. Let us begin here by understanding the very nature of an action, from a philosophical point of view.

We saw earlier that an action is not simply a behavior; actions are intentional. Coughing may be an action or it may be a mere behavior. I might cough to signal to someone that I want to interrupt her; but I can also cough reflexively due to an allergy to pollen. If the patient took or did not take the pill, then before doing it or not doing it she had the intention or did not have the intention to do it. As we shall see, the patient could also not take the pill because she did not intend to take it *or because she intended not to take it*. Intentions are thus partially constitutive of actions, and this makes it imperative to examine them. They are the very essence of the stage that precedes the act, and they confer on it the status of action.

Behind intention, there is motivation. On the difference between motive and intention, we may here quote Elizabeth Anscombe:

> To explain one's own actions by giving them a motive is to put them in a particular light. The question 'why?' often brings about this type of explanation [...]. The motives such as

admiration, curiosity, spite, friendship, fear, love of truth, hopelessness, and many others are either of this type, and very complex, or oriented-towards-the-future, or mixed. I say that a motive is 'oriented-towards-the-future' if it's an intention. For instance, to say that someone did something out of fear of... is often the same thing as to say that she did it out of fear that..., or in order that something does not happen (Anscombe 2000).

This is true even when the motivation is not connected directly to the action, but to a more general aim which the action helps to achieve. A patient will take or not take the pill, for example, depending on whether the motivation to do it or the motivation not to do it is *stronger*. But what do we mean by "stronger"? Are motivations one-dimensional, or do they have relative strengths and weaknesses? Is there a way to know a motivation's strength(s) other than by observing what actions actually result?

Before we know it, we find ourselves in very murky waters indeed. And for all our careful discernment of various mental contents, we have yet to touch on the most obvious phenomenon of all, that of the will. At what point in the cascade of events leading to an action does my will have a role? And how does my will intervene? First, it is difficult to know when a volition is supposed to occur (when did I really begin to want to do this, how long did it take me to want to do this, etc...); and, second, if volition were a stage of the action, it seems that it would itself require a stage of volition, and this antecedent would in turn require its own prior volition, etc. That is, the question seems to invite an infinite regression.

Our analysis of patient adherence seems to have led quickly to a vexing array of pernicious philosophical problems regarding volition, intentionality, and determinism. Lest we become even more lost, let us turn to the first modern theory of action, that of Donald Davidson. Davidson's work—and that of the many he has inspired—will be essential to our understanding of the paradox of patient nonadherence: Being motivated to be adherent yet not acting accordingly.

3.3.1 Davidson's Causal Theory of Action

Davidson proposed a causal analysis of action according to which an agent *does something because she believes* that, if she does it, what she *desires* will come to pass. For instance, I start to exercise because I have (*before* joining the gym) a reason to exercise (I want to lose weight) and because I believe that exercising is one of the actions that will make me lose weight (I also could go on a diet, take diet pills, etc). Although Davidson's model seems obvious and commonsensical, note how it differs from other models: Earlier behavioralist models stress the passivity of the person in that the person's behaviors are the almost mechanical product of various inputs. If the output is wrong (the patient is nonadherent to sound medical advice), then there must be something wrong with the inputs or with the person who converts those inputs to an output. Davidson, as we will see, has pried open the door of the black box.

According to Davidson, action is preceded by a 'pro-attitude'. A pro-attitude is simply a desire coupled with a belief that there is an action which can bring about the realization of that desire.

One can certainly imagine a great many desires, but of these, only a few will (I suspect) be obtainable. We do not form a pro-attitude regarding those desires; we desire to be the King of Spain, but we don't see this as within our reach, so no action results from this desire. On the other hand, I may desire a piece of toast, and I readily couple this desire with knowledge that putting a slice of bread in the toaster will lead towards satisfaction of this desire. Obviously, a pro-attitude by no means guarantees an outcome: I might find that I am out of bread, or that the toaster is broken. Or it might be that I may want a piece of toast, and have the means to make it, but still I do not do so because I am fasting prior to surgery.

Davidson argues that the state preceding an action fully explains that action. Again, this seems so commonsensical as to need no argument: But what Davidson is implying is that desires, beliefs, and other mental contents are not mere surface phenomena, underneath which are the "real" causal mechanisms (such as neuronal activity). To understand action, we must examine people at the level of their mental states, not their neurological states. This does not mean brain activity has no meaning—only that explanations based on neurological phenomena will be incomplete.

This model of action has yet another rich implication: Mental states and actions are deeply, but not inextricably, linked. I might believe that doing crossword puzzles will help me lose weight—it is a workout, I tell myself, albeit a mental one—but the results of my actions are not what I desired. Or, I find that doing crossword puzzles lead me to gain weight because I eat snacks at the same time. Or, I might start going to a gym to exercise, but binge eat afterwards. Or, I might not go to the gym, believing that working out will be too difficult (when in fact it probably wouldn't be so). Davidson's model of action allows for the normal things of everyday mental life: Mistaken notions, misunderstandings, unintended consequences.

Much depends on what a person *sets out* to do: To understand human action, it matters critically what a person is attempting to accomplish. Surmising the person's desires merely on the basis of her behavior is perilous, just as desire does not determine behavior either.

In Davidson's conception, a desire-belief pair has two roles. First, it explains the action and, second it allows us to rationalize the action. It is because I want to lose weight and because I believe that exercising can help me to lose weight that I start doing it. This is the reason I could give if someone asked me: Why did you start exercising? Davidson goes one dramatic step further: *He contends that this reason is also the real cause of the action* (Fig. 3.1) (Davidson 2001, 3–20).

Here Davidson defied his contemporaries and the philosophical tradition. Wittgenstein, most notably, stopped at the explanatory role of reasons of actions and denied the possibility that a mental event could causally influence a physical event. Davidson notes that if we are to understand our everyday notion that we act because of some particular reason, we must acknowledge that a mental event causes a physical action—a concept anathema in philosophy and science, despite its implicit acceptance in everyone's life (including philosophers and scientists!). Davidson:

> In order to turn the first 'and' to 'because' in 'He exercised and he wanted to reduce and thought exercise would do it', we must, as the basic move [admit that] a primary reason for an action is its cause (Davidson 2001, 9–12).

Fig. 3.1 Davidson's causal
theory of action

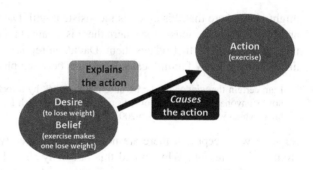

We are now justified in putting a causal arrow between the mental states and the physical phenomenon (Fig. 3.1).

The cause for an action does not, of course, determine what the outcome will be; nor does each action have just one cause. For instance, I could start exercising for a completely different reason; I might start exercising in order to meet someone who also goes to this gym. In this case, it is not the desire to lose weight that *causes* me to exercise. In addition, by limiting his description of an action to what the agent actually does and the reasons for doing it, Davidson refuses to include the results of an action in its description. A driver who jams on the brakes to avoid hitting the car in front has indeed performed an action, whether or not he succeeds in preventing a collision. Intentions matter, as any attorney knows: A bullet fired in anger has far different implications than a bullet which is fired by accident, even if a person is killed in each case. Outcomes are not irrelevant in the law, but they are understood in the context of the intentions of the actor.[2]

Indeed, Davidson concludes that *primitive* actions, i.e., the ones that the agent actually performs, are the only actions that exist, and that as far as the *effects* of one's actions are concerned, "the rest is up to nature" (Davidson 2001, 53). When I flip a light switch, illuminating the room and alerting the burglar, only flipping the light switch counts as my action.

Davidson does not deny that "an agent causes what his actions cause" (Davidson 2001, 59) rather, he maintains a sharp distinction between an agent's actions and their causal effects. I joined and exercised at a gym; these were indeed my actions. And though we may often *say*, in addition, that I lost weight, losing weight was in fact the result of my action, not an action I in fact performed.

There is a real causal relationship between mental states (desires and beliefs) and physical events (an action). And yet we cannot say that there is a law that states that when one has such and such desire and such and such belief, a certain action will necessarily result (in reality, I still have not joined a gym). One

---

[2] An exception which proves the rule is the existence of strict liability laws: In such cases, intentions are explicitly excluded from consideration of culpability. Only the outcome is relevant in strict liability cases, for example, making false statements on a loan application. I am grateful to John Meyers for this remark.

might object here that this claim is inconsistent with Davidson's claim that desires and beliefs cause actions—for where there is a causal relation between two events, there must be a law that relates them. Davidson replies that the relation here is no different from many familiar causal relations between physical events:

> I am certain the window broke because it was struck by a rock – I saw it happen; I am not (is anyone?) in command of laws on the basis of which I can predict what blows will break which windows (Davidson 2001, 16).

Suppose we accept that there are no laws which specify which action will result from a given reason: Why should this be? Why should reasons be causative but not determinative? One explanation of this is as follows: Beliefs and desires bring about action only when a wide range of other conditions are held constant. That is, actions always occur in an environment which powerfully shapes the outcome. For example, a desire to stay dry and a belief that it's raining typically causes me to bring an umbrella on my walk to work; but not if I don't believe that rain can get me wet, or if I believe umbrellas are ineffective for keeping me dry, or if I think the clouds will part as soon as I leave home, etc. These implicit beliefs shape whether my desire to stay dry will cause me to pick my umbrella up.

Not only do these implicit, background beliefs shape the outcome of an action, we also routinely alter our environment to make generally desired outcomes possible. For example, I may keep an umbrella near the front door so that when I desire to stay dry on a rainy day, I actually have an umbrella at hand when I'm leaving. The environmental regularities we establish are usually unseen—until they are disrupted by moving to a new home, a different city, or the like.

We have long been mistaken in thinking that there should be a law connecting reasons to actions owing to the ancient idea that actions are the result of practical syllogisms which mirror the form of the theoretical syllogism. From Aristotle on, the logic of action was understood to follow the form of a *practical syllogism*, a deduction from a major premise (all men are mortal) and a minor premise (Socrates is a man), to a conclusion (therefore Socrates is mortal). Similarly, an action could be seen as consisting of (1) of a major premise, a desire (loosing weight: All actions that make me lose weight are desirable), and (2) a minor premise, a belief (I believe that exercise is this type of action), from which follows (3) a conclusion. The action, then, follows immediately (so, immediately, I start exercising). Note that the action is supposed to follow from the conclusion *immediately*, just as the conclusion of a theoretical syllogism is supposed to do, without intermediaries. This was the syllogistic, deductive, conception of action: A reason explains an action by virtue of their logical relationship as premises to conclusion.

Ruwen Ogien (1993) has argued that a theoretical syllogism and a practical syllogism differ in two important ways. First, in the case of a practical syllogism, the major premise is not simply a declaration such as "all men are mortal", which is a fact. Rather, the premise of a practical syllogism expresses a wish, a desire or an intention; it is not so much a statement of fact as of possibility. If, for instance, I declare "All men are mortal" and am immediately struck dead by lightning, the truth that "all men are mortal" doesn't change one iota. On the other hand, if I say "I want to eat apple pie", the "truth" of this assertion must still play out (do I actually eat pie when offered? Do I refrain, but explain I'm fasting that day? etc.) Being struck immediately dead puts a quick end to knowing the truth of my statement in a way that isn't so for the statement that "all men are mortal." Statements of fact and statements of intention and desire, therefore, are different.

Second, the major premise of a practical syllogism does not have a categorical, 'universal', aspect: According to Ogien,

> Let us suppose the first major premise, the desire 'I want to eat something sweet: and the minor premise, the belief: 'This orange is sweet'. The major premise cannot have a universal aspect that would make it so that each time something is sweet, I eat it. Because this would imply that I eat every orange that I believe is sweet, even if I also believe, for example, that this particular orange was poisoned. Does this imply that my major premise loses all its practical strength and that my desire to eat something sweet can never lead to an action? Indeed, a syllogism can have a practical strength only if its major premise is unconditional or categorical, in one way or another.

A practical syllogism begins, then, with belief or desire which is necessarily contingent. It is only true in certain conditions. "What goes up must come down" is true only if we also assume that it goes up at less than escape velocity, that we are in a gravitational field and not in outer space, and so forth. The practical syllogism is based on a major premise which has various "ifs", "buts", and "excepts in the case of" appended to it.

Davidson argues that a practical syllogism—composed as it is of contingent premises—cannot follow the same deductive logic of the theoretical syllogism. Ogien notes:

> We can now understand why Davidson believes that he has radically turned away from the supposed practical syllogism. Since the action is no longer *deduced* from a conditional universal proposition; it is an event that is *put in relation* with more or less reasonable beliefs (Ogien 1993 54–65).

To return, then, to our original quandary: Why don't reasons always and necessarily result in the actions that they entail as a matter of logic? Now we can respond: The causal theory of action breaks the bond between the end of the deliberation and the action itself–it proposes that the 'reason' for the action (the desire-belief pair) has two roles: In addition to its explanatory role (that can be given *a posteriori*), it has a truly causal effect. But this causal effect cannot occur until the agent, to use Davidson's term, 'has considered all things' (Fig. 3.2).

The phrase 'all things considered' does not mean "a consideration of all conceivable things" but rather refers only to a consideration of things known, believed, or held by the agent, the sum of his relevant principles, opinions, attitudes, and desires (Davidson 2001, 40).

The realization of an action is connected, therefore, to a set of pertinent mental states and the holistic organization from which those mental states arise, as well as factors external to the person. This concept of action eschews deterministic laws, but does acknowledge the force of mechanisms—mechanisms which are employed afterwards to explain an action, and to predict the likelihood of future action when similar circumstances obtain.

Fig. 3.2 Acting 'having all well considered', a simplified example

Davidson's philosophy is described by the term anomalous monism, meaning there is a unity of body and mind (monism), but the relationships of mental events are not regulated by laws (anomalous).

Can we apply the abstract ideas developed in this chapter to the mundane problem of patient adherence? In the next chapter, we will try to do just that: To re-analyze adherence (and nonadherence) from a perspective of action, rather than behavior. In studying the patient as an agent—rather than as a performer of behaviors—we will develop the concept of 'therapeutic agency'.

References

Anscombe GEM. Intention. 2nd ed. Cambridge: Harvard University Press; 2000.

Baumeister RF, Vohs KD, DeWall CN, Zhang L. How emotion shapes behaviour: feedback, anticipation, and reflexion, rather than direct causation. Pers Soc Psychol Rev. 2007;11:167–203.

Bower GH, Forgas JP. Affect, memory and social cognition. In: Kihlstrom JF, Bower GH, Forgas JP, Niedenthal PM, editors. Cognition and emotion counterpoints: cognition, memory and language. Oxford: Oxford University Press; 2000.

Clore GL, Gasper K. Feeling is believing: some affective influences on belief. In: Fridja NH, Manstead ASR, Bem S, editors. Emotions and beliefs, how feelings influence thoughts. Cambridge: Cambridge University Press; 2000.

Davidson D. Actions, reasons and causes. J Philos 1963; 60: 685–700. Reprinted In: Essays on actions and events. Oxford: Clarendon Press; 2001.

Davidson D. Two paradoxes of irrationality. In: Wollheim R, Hopkins J. editors. Philosophical essays on Freud. Cambridge: Cambridge University Press; 1982. p. 289–305. Reprinted In: Problems of rationality. Oxford: Clarendon Press; 2004.

De Souza R. The rationality of emotion. Cambridge: MIT Press; 1987.

Descombes V. Les Institutions du sens, Éditions de Minuit, 1996.

Dokic J, Engel P. Ramsey, Vérité et succès, P.U.F., collection Philosophies; 2001.

Elster J. Getting hooked, rationality and addictions. In: Elster J, Skog O-J, editors. Cambridge: Cambridge University Press; 1999a.

Elster J. Alchemies of the mind. Cambridge: Cambridge University Press; 1999b.

Elster J. Strong feelings, emotion, addiction and human behavior. The 1997 Jean Nicod lectures, A Bradford book. Cambridge: The MIT Press; 2000.

Engel P. Les croyances. In: Notions de Philosophie, under the direction of D. Kamboucher, Éditions folio, Gallimard; 1995.

Engel P, editor. Believing and accepting. Philosophical studies series. Dordrecht: Kluwer Academic Publishers; 2000.

Fridja NH, Mesquita B. Beliefs through emotions. In: Fridja NH, Manstead ASR, Bem S, editors. Emotions and beliefs, how feelings influence thoughts. Cambridge: Cambridge University Press; 2000.

Heil J. Philosophy of mind, a contemporary introduction. London: Routledge; 1998.

Hume D. A treatise of human nature, 1.3.7; 1740.

Leibniz GW. Monadology, 1698.

Livet P. Émotions et rationalité morale. P. U. F., Collection Sociologies; 2002.

Loewenstein GF, Weber EU, Hsee CK, Welch N. Risk as feelings. Psychol Bull. 2001;127:267–86.

Losonsky M, Jean-Pierre Dupuy JP. In: Engel P, editor. Believing and accepting. Philosophical studies series. Dordrecht: Kluwer Academic Publishers; 2000.

Maslin KT. An introduction to the philosophy of mind. Cambridge: Polity Press; 2001.

Nozick R. The nature of rationality. Princeton: Princeton University Press; 1993.

Ogien R. La Faiblesse de la volonté. P.U.F.; 1993.

Pears D. The goals and strategies of self-deception. In: Elster J, editor. The multiple self. Cambridge: Cambridge University Press; 1985.

Pharo P. Raison et civilisation, essai sur les chances de rationalisation morale de la société. Cerf; 2006.

Ramsey F. Philosophical papers. In: Mellor DH, editor. Cambridge: Cambridge University Press; 1990.

Russell B. Human society in ethics and politics. London: Allen and Unwin; 1954.

Searle J. Intentionality, an essay in the philosophy of mind. Cambridge: Cambridge University Press; 1983.

Tykocinski OE, Pittman TS, Tuttle ES. Inaction inertia: foregoing future benefits as a result of an initial failure to act. J Pers Soc Psychol. 1995;8:793–803.

Wittgenstein L. On certainty, 141, 142; 1951–1952, published 1969. http://evans-experientialism. freewebspace.com/wittgenstein03.htm. Accessed 7 May 2013.

Chapter 4
An Intentionalist Model of Patient Adherence

Abstract If we understand patient adherence as to the performance of a therapeutic action, we can now propose an 'intentionalist' model of adherence, establishing a certain hierarchy of mental states in which emotions play as vital a role as knowledge, beliefs, skills, etc. In this model, emotions play this role in inducing revisions of beliefs, expectations, and preferences of patients' various desires. It is obvious that non-intentional factors (such as pain or pleasure), can have a motivational role—indeed, sometimes overwhelmingly so. Events can intervene as a substratum of new beliefs, or by provoking the emergence of emotions. Exogenous factors, such as the presence or absence of resources can intervene in encouraging or limiting patient adherence. It is important to recognize that a fundamental gap exists between therapeutic actions and their results. For not only do actions not assure a desired result, they may only play a partial role if the desired result does happen. At best, our actions are only partly responsible for what happens to us. For example, a patient may increase her insulin dose to reduce her blood sugar after dinner. But it is misleading to say that her action was to 'normalize her blood sugar'. This does not mean, of course, that by adjusting her insulin dose, the patient does not try to normalize her blood sugar (and that her doctor must not do everything in order to convince her to try to do it).

4.1 Therapeutic Agency

Right away, this concept of 'therapeutic agency' throws a new light on many common situations of adherence or nonadherence. Jane's belief that insulin makes one gain weight (even though it is questionable) makes her inclined to perform actions which become now understandable: When paired with, say, a desire to lose weight, it inclines her to lower her insulin doses or even to refuse the insulin treatment altogether. Tom wants to please his diabetologist; paired with the belief that adjusting his insulin doses will make his doctor happy, he is inclined to do so. John *wants* to take care of his health and that means that John is inclined to take

© Springer International Publishing Switzerland 2015

G. Reach, *The Mental Mechanisms of Patient Adherence to Long-Term Therapies*,
Philosophy and Medicine 118, DOI 10.1007/978-3-319-12265-6_4

55

an aspirin *because he believes* that this gesture is among the actions that contribute to better health.

But the human mind is more complex: We saw that the result of an action could modify the content of the propositional attitudes leading up to it. Consider a patient who, having adjusted her insulin dose, suffers a serious and frightening hypoglycemic episode during which she loses consciousness. This will likely make her skittish—at least—the next time she considers taking additional insulin. She might, for example, not increase her insulin (though she might need to), or she might choose to take more (though she really shouldn't). Our actions become lessons which modify our beliefs, which in turn lead to different actions in the future.

The concept of therapeutic agency appeals directly to what a patient thinks of the therapeutic act (her desires and beliefs), rather than simply her knowledge or her hypothetical feeling of obligation or duty: It recognizes a gap between knowledge and belief, and between duty and what one actually does (or doesn't) do. Although one may vote out of civic duty, such motivations toward adherence are rarely decisive. The notions of obligation and duty usually do not play a role here. Sometimes one hears the argument that there is an actual, almost civic duty to take care of one's health (for instance to help the State avoid the healthcare costs that would be the consequences of my negligence): This is unlikely to motivate many persons.

According to the causal theory of action, an action—for our current purposes—is better characterized by its reason than by its result. Adherent actions do not guarantee salubrious results. The disappearance of a painful bladder infection depends crucially on the natural world's cooperation: How the offending bacteria respond to the antibiotic, for example. She may adhere admirably to the treatment and still not get better.

Indeed, a fundamental gap exists between actions, no matter how well-informed, and the results. For not only do actions not assure a desired result, they may only play a partial role if the desired result does happen. At best, our actions are only partly responsible for what happens to us.

For example, a patient may increase her insulin dose to reduce her blood sugar after dinner. But it is misleading to say that her action was to 'normalize her blood sugar'. Indeed, we should not confuse an objective standard that physicians agree on (patients should have a blood sugar value within normal limits) with a goal to be given to the patient (bring your blood sugar within normal limits). Rather, one must realize that her blood sugar does not depend entirely on her behavior (Wolpert and Anderson 2004); countless other factors–some known, some unknown, and some unknowable—have a role to play. To say that someone normalized her blood sugar is just as absurd as it would be to say that she intentionally had a heart attack.

This has important implications: We can understand that numerous studies attempting to find a correlation between, for instance, educational programs and a result, or a consequence, such as the level of glycated hemoglobin in the treatment of diabetes have negative results. It is because the dependent variable is too far from the independent variable. Perhaps it would be more revealing to study the correlation between an educational program and the tasks actually performed by patients (insulin injections, measuring the blood glucose, adjusting the doses of insulin, etc.) We saw earlier that it is difficult to prove that programs aimed to improve patient adherence are effective. In most of them, glycated hemoglobin is used as the primary endpoint.

Similarly, it would be absurd to ask an obese patient to reach her 'ideal weight', even if physicians or the insurance companies (who have defined what "ideal" is) think it offers the highest chances of survival.

This does not mean, of course, that by adjusting her insulin dose, the patient does not *try* to normalize her blood sugar (and that her doctor must not do everything in order to convince her to try to do it). This can be compared to the fundamental problem of the way we perceive the concept of causality, as it was analyzed by John Searle. The patient increases her insulin dose in order to lower her blood sugar, because she was taught, and maybe especially because she noticed through experience, that there is a certain regular correlation between increasing the insulin dose and the lowering of the blood sugar. In the 'Background' (as Searle would say) she has the idea of a regularity between what events in the world. This very notion of regularity authorizes the concept of learning through education or experience. Searle gives the example of a basketball player:

> When I try to shoot free throws from the free-throw line I am only occasionally successful. But the point is that, when I do succeed, things go according to plan (Searle 1983, 138).

In the same vein, this does not mean that by staying on a diet, the obese patient does not *try* to reach the indicated weight. This is exactly what John Searle says:

> If I intend to weigh 160 pounds by Christmas and I succeed, it won't do to say I performed the intentional action of weighing 160 pounds by Christmas nor will it do to say that weighing 160 pounds by Christmas can be an intentional action. What one wants to say rather is that if I fulfilled my intention to weigh 160 pounds by Christmas, I must have performed certain actions by means of which I came to weigh 160 pounds (Searle 1983, 80).

As Davidson stresses:

> Trying to do one thing may be simply doing another. [...] It is this fact too that explains why we may be limited, in our actions, to mere movements of our bodies, and yet may be capable of [...] from time to time, hitting the bull's eye (Davidson 2001, 60).

The Medication Event Monitoring Systems, MEMS, measure patient adherence by counting each single event, i.e. opening a pill bottle, while ignoring the result of the act (the patient may open the bottle and put the pill under her bed). This is doubtless one of the best illustrations of the fact that patient adherence can be viewed as the acceptance to perform repeated actions, in Davidson's sense of the word.

4.1.1 To Take Care of Oneself or Not

Is not taking a pill (the occurrence of *non-a*) the same as the absence of taking a pill (the non-occurrence of *a*)? If we focus only on the result (*p*), there should be no difference. But in our analysis of human agency, we find that they are not the same at all. Willful refusal to take a medication and simply forgetting to take that medication have very different implications, and calls for different actions on the part of the physician.

Let's analyze this according to Davidson's schema of *primary reasons*. For the action of taking a pill, the primary reason might be: "*I see something desirable in all actions* that will make my bladder infection disappear; and *I believe that* taking this pill is this type of action." For the action of not taking the pill, a primary reason might be: "*I see something desirable in all actions* that prevent the diarrhea that I previously had because of this antibiotic; and *I believe that* not taking this pill is this type of action."

It is immediately clear that the desires (i.e., the causes) of the two actions (to take or not to take the pill) are different, and that neither is simply the opposite of the other. So there is a difference between taking and not taking a pill that goes far beyond one being the negative of the other, even if they share an important common feature, namely, that the medication is not ingested.

In other words, the non-occurrence of *a* is not the same thing as the occurrence of *non-a*, not because of the result *p*, which is the same, but because of the difference in reasons, i.e. the causes, of *a* and of *non-a*.

4.2   An Intentionalist Model of Adherence

If we understand, as I have proposed to do in this book, patient adherence as to the performance of a therapeutic action, we can now propose an 'intentionalist' model, establishing a certain hierarchy of mental states in which emotions play as vital a role as knowledge, belief, skills, etc. (Fig. 4.1).

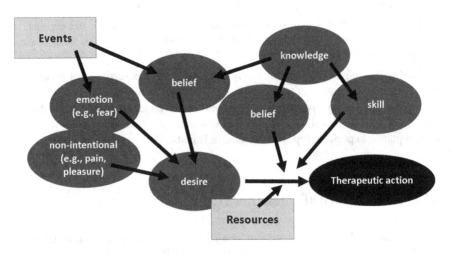

Fig. 4.1   An intentionalist model of adherence

Schematically, desire is the pro-attitude that leads to an action: For instance, Jane wants a beautiful baby, so she begins to adjust her insulin dose correctly. Beliefs play an instrumental role: She does it because she believes that correctly adjusting her insulin dose will make it more likely that her baby is beautiful (she had the same belief before she was pregnant, but it didn't motivate her to adjust her insulin dose). Desire can also be caused by another belief. For instance, Irene believes that she will be happier if she weighs 10 pounds less, and this causes her desire to lose weight. Similarly, an emotion can cause a new desire: Thomas wants to take care of himself because he fears complications. Skill also plays an instrumental role in action. Jane must *know how* to adjust her insulin doses, although this is not *why* she does it.

In this model, emotions play a major role in inducing revisions of beliefs, expectations, and preferences of patients' various desires. It is obvious that non-intentional factors (contentless states), such as pain or pleasure, can have a motivational role—indeed, sometimes overwhelmingly so. Events can also intervene as a substratum of new beliefs, or by provoking the emergence of emotions. Exogenous factors, such as the presence or absence of resources, for example, can intervene in encouraging or limiting patient adherence. Finally, to follow Searle, the different mental states listed here are inscribed in a Background of competence and presuppositions (not represented in this graphic) which allows those mental states to 'function'. One notes that patients' desires are a driving force in this model (it is clear that the modules representing, for instance, beliefs and desires, must be seen as generic. In particular, the so-called secondary order mental states certainly play a very important role: Beliefs about beliefs, desires about desires. I will come back later to this important concept).

Events, mental states, and exogenous factors can each have a positive or a negative effect on patient adherence. Some factors may influence actions in a predictable way, but have the opposite effect in some patients, or at some times. For example, physicians know that a medication given once a day is more likely to be taken as prescribed than a medication given four times a day—yet a few patients want higher frequency dosing, and respond better on such a regimen.

4.3 The Pivotal Role of Emotions in Patient Adherence

Emotions were barely present in the various models of adoption of a health behavior described in the first part of this book (Leventhal's Self-Regulation Model is the exception). Leventhal explicitly suggests that a patient's evaluation of her own behavior is partly cognitive and partly emotional (Leventhal et al. 1997). In the other models, emotions are seen as a nuisance which divert a "susceptible" patient from the path of rational treatment adherence.

However, just what do we mean really by emotion? For our purposes here, we may think of emotions as affective (i.e., felt) states which are *about* something. Emotions may be distinguished from visceral states—which are certainly felt—but which are not directed at anything (they are non-intentional, to use our earlier language). Elster gives the following groupings:

> Among the states that unambiguously qualify as emotions we may first list various social emotions: anger, hatred, guilt, shame, pride, pridefulness, admiration, and liking. Second there are various counterfactual emotions generated by thoughts about what might have happened but didn't: regret, rejoicing, disappointment, elation....Third, there are emotions generated by the thought of what may happen: fear and hope. Fourth, there are emotions generated by good or bad things that have happened: joy and grief. Fifth, there are emotions triggered by the thought of the possessions of others: Envy, malice, indignation, and jealousy. Finally, there are cases that do not fall neatly into any special category, such as contempt, disgust, and romantic love...borderline or controversial cases include surprise, boredom, interest, sexual desire, enjoyment, worry, and frustration (Elster 1998).

Pierre Livet defines emotion somewhat differently, emphasizing the generative role played by the never-ending flux of experience:

> [Emotion is the] affective, physiological and behavioral resonance of a differential between one or two perceived (or imagined, thought of) traits of the situation and the continuation of our thoughts, imaginings, perceptions or actions currently under way (Livet 2002, 23).

Livet implies that, as with the five senses, emotions are driven by contrast. But unlike the senses, which are more tightly grounded in the present, emotions may arise from a variety of differences: The present versus the anticipated future; the present versus the past; one's self-perception versus the perception of others; desire versus what the environment has to offer; and so forth. Each of these differences—imbalances, if you will—gives rise to emotional states. These states may be transient, enduring, or even unnoticed (skilled psychotherapists, for example, are adept at picking up transient/ignored emotions as a way to understand the patient's trouble, for example). Except for strongly felt emotions, or ones which are enduring, much of our emotional life is seamlessly experienced—just as our sensory experience is.

Livet's concept of emotion enables us to understand how the announcement of a chronic illness (even if it were expected and no great surprise), generates diverse emotions. It is the difference between the immediate past and the new—very new—present which produces the fear, anger, anxiety, dread, regret, guilt, disgust, even relief, felt by patients who are diagnosed with a long-term illness.

On first examination, Livet's idea may seem to be mere common sense dressed in fine language. But as with Davidson, it is the ramifications of the idea which reveal its profundity. If emotion is about difference, it means that emotions never arise *sui generis*: Every emotion has an origin. And because the differences which give rise to emotions are so diverse, it helps us to understand how someone experiences conflicting emotions. In a purely rational model of human experience, conflicting emotions are not supposed to occur—how can one person be of two minds? But Livet opens the door to explaining something everyone experiences: We have conflicting emotions when we have multiple differences in play. For example, a patient newly diagnosed with a chronic illness may feel guilt ("I knew I wasn't taking good care of my health, and now I'm being punished") and relief

("I was having mysterious pains before, and now at least I know what's causing them."). Relief may then give way to fear ("I was relieved to know what ails me, but now what's going to happen?"); and so forth.

Livet also helps us understand both the fragmentary nature of our experience—differences arise and fade away all the time—and how such fragments nevertheless form into a coherent experience of our life. Because emotions never arise from a vacuum, they remain connected to previous emotions and experiences. Over time, patterns recur—fostered in part by the many regularities in our physical and social environments—and these form enduring elements in our mental world. It is not unlike the puzzle metaphor discussed earlier.

Emotions are by no means passive responses to "difference." Emotional states motivate action, often in fascinating ways. In the simplest cases, an emotion such as envy might lead a person to take a friend's coveted possession; or anger may lead to aggressive behavior. However, that same envy may lead to avoidance of that friend, lest the envious party be constantly reminded of what she doesn't have. Or, envy may lead to an attempt to not desire that which is envied: If the desired "possession" is, say, a boyfriend, the envious friend may dwell on his deficiencies to make him seem less desirable. Or, even more complexly, she might make a sexual advance towards her friend's boyfriend, so she can imperiously reject him if he reciprocates, or—if he doesn't—she can dismiss him as unmanly.

What is important to understand is that our emotional lives are complicated in a way which is part and parcel of being human. Our attempt to better understand adherence will necessarily involve the patient's emotions, and all the actions driven by them. If our goal is to understand a real world issue—rather than a theoretical one—then emotions must be given their proper place in explaining human action.

When it comes to making treatment decisions, emotions play an especially big role, as illness and health are of existential concern to the patient. In anger, Juliet comes to believe that the doctor doesn't know anything—he's a complete moron. The truth that John has cancer motivates him to believe the very opposite, that he is in perfect health. John is so afraid of being sick that, despite all available evidence (the weight he has lost, the X-rays and the lab exams that he has seen), he believes that he is healthy. The announcement of one's illness can be scary; and so it is no surprise, then, that it can, on occasion, result in denial and unhealthy behaviors.

If we accept that emotions are the result of a changing state of affairs—an imbalance—then perhaps it makes sense that this imbalance is sometimes too much to handle, and can lead to overcompensations or other drastic actions to right the suddenly listing ship.

4.3.1 Emotions, Boredom and Anxiety

Pierre Livet's interpretations of boredom and anxiety are also pertinent here. Boredom is the *opposite* of what he calls 'being used to'. When we are simply "used to" some state of affairs, due to the revision process, emotions disappear

because the differential between reality and our perception of the world diminishes. In the case of boredom, Livet notes that the world does not change, and

> not causing any revisions because we will have accomplished the necessary revision, it
> does not cause any revisions when we were expecting suggestions of revisions. The world
> no longer produces the basic differential that gets in the way of the normal course of
> events (Livet 2002, 32).

According to this conception, the patient experiences no differential, and therefore no more emotion. This helps explain how exasperatingly dull the treatment of
a chronic illness can be: One of its most grinding characteristics is that it requires
indefinite treatment. This may explain why patients pay so much attention to the progress in medical research perhaps because a "medical breakthrough" is like a floata-
tion ring to the floundering swimmer—and not just because it can save a patient's
life, but because it might relieve the crushing tedium of treatment without end.

As for anxiety, Livet sees it as an emotion "secondary to all emotions, when the
revision [that it requires] turns out to be difficult and uncertain." (Livet 2002, 31)
Anxiety, then, is felt when an emotion-driving difference is not—or cannot—be
worked through. It may be that the resolution is delayed (as with waiting to take
an exam at school next week), difficult (undergoing a painful medical procedure),
or where the outcome is not clear (going down a ski slope and not knowing if one
is skilled enough to not crash).

We shall see by the end of this book that patient nonadherence is in a way due
to the inability to fully engage in the necessary revision of beliefs, expectations
and preferences which leads to the acceptance of treatment. The preceding sections
help explain how being unable to make these crucial revisions leads to the persistence of the emotions which accompanied the start of the illness, such as fear,
anger, guilt, disgust, shame, and the like. When patients cannot move forward, they
indeed stay in the same place—as, for instance, the alcoholic who "does the same
foolish thing over and over, expecting different results," or the diabetic patient who
is repeatedly careless about his blood sugars and is in the emergency room every
few weeks. Not only does this repetition of ineffectual action imperil the patient,
it is a common source of frustration and bewilderment to friends and family, who
just cannot fathom why their loved one cannot see her way out of this spiral.

Anxiety—so commonly seen in patients who have not accepted their illness and
its treatment—is also a product of this "stuck" state. This anxiety is in itself distressing, and may lead to denial of being ill. The psychological maneuver of denial
is, like many such defenses, effective in the short-term, but debilitating in the long
run. When a patient stops living in a state of denial, she experiences a sense of
relief and "being present", despite now consciously facing her illness squarely.

4.3.2 Emotions and Patient Adherence

Emotions are, not surprisingly, a factor throughout an illness—not just at the time
of diagnosis. For example, a study shows that cancer patients are more likely to

opt for a treatment which is presented in terms of chance of survival rather than one presented in terms of chance of death, even if the mortality rates of both treatments are the same. The results were identical among patients, medical students and physicians, suggesting that the emotions associated with terms for life and death intervened in the decision making process, independent of the agent's medical (McNeil et al. 1982).

However, emotions can play as well a positive role in patient adherence. The fear of future complications of the disease, the announcement of the occurrence of the first complication, or the death of a relative from the same illness can powerfully influence whether a patient accepts treatment. Among patients with diabetes, the first instance of a retinal microaneurysm (in itself an unimportant complication) is sometimes a turning point for the patient: The news of a complication transforms the illness from an abstract concern to a concrete worry. Some public health campaigns appeal to fear in the fight against smoking or drunk driving. This use of fear as a motivator is not straightforward: Young smokers might be shown a picture of a person with advanced lung cancer to demonstrate the fate which awaits them; yet few would condone showing the same pictures to persons newly diagnosed with cancer (instead, they might be given a motivational speech about how they can still live a full life, etc.) In short, it matters very much where a person is in the progression of an illness and what their pre-existing motivations are.

In the satirical play (1923) Dr. Knock, *or The Triumph of Medicine* (Romains 1972), Dr. Knock instructs his patient to imagine a crab, an octopus, or a giant spider nibbling at her and tearing her brain to pieces. In response, she slumps in an armchair—it's enough to make one faint from horror, and she says:

> O! I'll be a very docile patient, Doctor, like a little dog. I'll do everything, especially if it's not too painful (Romains 1972, 99–100).

"Positive" emotions have a role to play, of course. One study is particularly revealing: Its goal was to assess the effectiveness of health education in the treatment of obesity among Pima Indians of North America. Pima Indians were formerly hunters in the areas of Arizona and northern Mexico; as such they experienced the "boom or bust" eating pattern of hunters characterized by feasting on days when a kill was made, interspersed with periods of mild starvation when the hunters were unsuccessful. Persons with this dietary pattern secrete insulin slowly after meals, which results in a higher overall secretion of insulin and better storage of excess calories in the form of fat. It was advantageous to have some fat to carry one through the lean times; one could not be certain on any given day of finding sufficient food. The randomness—not just the scarceness—of the meals had a regulatory effect on weight. But with the eventual coming of farming in the 19th century and the steady diet it provided, Pima Indians were deprived of this regulatory factor.

Obesity and diabetes among the Pima has increased dramatically in the last few decades, foreshadowing what is now happening worldwide. Narayan and his colleagues randomly exposed two groups of Pima Indians to two education programs. One group was taught the major principles of nutrition; the other was taught about their civilization and history. Surprisingly, it is only the second group, called the '*pride* group', that showed positive effects in terms of weight loss and improvement of diabetes. Obesity and glycemic equilibrium actually deteriorated more in the group given nutritional education than in the group of Pima who declined to participate in the study (Narayan et al. 1998).

4.4 Bringing Action into Play: Volition

Cognitive psychology has neglected the study of human will over the past several decades. Rather, psychologists have spoken of decisional processes, executive functions, volitional capacity, conditioned responses, and the like. In her book on the nature of the will, the philosopher Joelle Proust reminds us of John Locke's description of will and volition. Though written well over 300 years ago, Locke's will is quite in line with our own commonsense ideas about what it is:

> We find in ourselves a power to begin or forbear, continue or end several actions of our minds, and motions of our bodies, barely by a thought or preference of the mind ordering, or as it were commanding, the doing or not doing of [some] particular action. This power which the mind has thus to order the consideration of any idea, or the forbearing to consider it; or to prefer the motion of any part of the body to its rest, and vice versa, in any particular instance, is that which we call the Will. The actual exercise of that power, by directing any particular action, or its forbearance, is that which we call volition or willing. (John Locke, *An Essay Concerning Human Understanding*, Book 2, Chap. 21, 1690).

For Joelle Proust, volition is

> a basic action: volition is the event through which the agent puts herself in a position to act with a goal in mind (…) and the complete action, the one that includes the change in the world, that is the goal of the action, is the causal expansion of this basic action (Proust 2005, 161).

Preceding the volitional act, a person must:

(1) have a representation of the means allowing her to achieve this goal (she has already done it in the past in a given motivational context): This explains how a past occurrence allows an organism to reproduce an effect in a new, inevitably different situation.

That is, a willful act is more than the output of a behavior; to be willful, an act must have some degree of abstract generality. Willful acts are malleable, to greater or lesser extent, and when an action lacks this quality, we notice it right away—consider the epileptic having a seizure or the sleepwalker wandering about the house.[1]

---

[1] This aspect of willful action is important in criminal law—for acts which lack willfulness are not punishable. Most legal definitions of insanity, for example, include the idea that the insane person was unable to conform his behavior to the requirements of the law, so that regardless of the damage wrought by the behavior, the insane defendant is not held responsible. The criminal law struggles with liminal cases such intoxication, delusional insanity, actions committed by sleepwalkers, and so forth, where a great deal hinges on whether an action was willful or not. I am grateful to John Meyers for this comment.

Furthermore, Joëlle Proust notes that part of the set up for a volitional action is that

(2) the context again presents the achievement of the goal as attractive ('salient'), in regard to the new needs of the agent: this clause stresses the importance, in this model of action, of the cyclical nature of needs and desires. (3) Finally, that the agent is able to put herself in the situation where she is able to make the effort that corresponds to the action in question, whether she feels she is capable of it, whether she foresees being able to bring the action to its conclusion, etc. (Proust 2005, 138).

Proust's model gives us an important key to understanding patient adherence. In particular, this model of volitional action incorporates a stage in which the person makes herself capable of acting—a step in which she readies herself to do something. It accounts for the many actions we abort each day which, precisely because we *don't* do them, tend to be ignored. It is not only describing restraint—I don't touch the hot stove, I don't light up a cigarette—but also delay, reconsideration, equivocation and hesitation.

When we act in a certain way, there are many ways we might otherwise have acted. In crossing a busy street, for example, we might walk across, or dart quickly; we could also decide to not cross at all, or to wait until we got to an intersection with a light. (We are only mentioning those options which might reasonably be considered. We could cross with our eyes closed, or while doing backflips, etc. What is important is not all conceivable actions, but the ones which we give plausible consideration to). In crossing the street, we might, for example, decide to cross later; in such a case, have we decided to cross the street, or not? Or, perhaps being a bit fearful of all the traffic—but really needing to get to the other side—we might take a step off the curb during a promising gap, only to jump back when we realize we can't make it. Such "toe in the water" actions are well accounted for in Proust's schema.

Proust's model also allows us to understand better the relationship between pleasure and desire (see Fig. 4.1). We saw earlier that we could—following Patrick Pharo—define pleasure as "an experience that one does not get tired of, or that one gets tired of only when it no longer gives pleasure, i.e., when it is no longer a pleasure". As long as one does not become tired of a pleasurable experience, it remains vivid and can come to be identified with the representation of the goal of the action (condition 1); and, it is part of the salient character of the goal, which will again be presented to the agent (condition 2). In short, we are drawn to pleasure, sometimes to the extent where pleasure becomes an end in itself. Not all things we desire are pleasurable, but all that is pleasurable we desire (though we may have various reasons for delaying or denying ourselves that pleasure).

Further, Proust's description of the cyclic nature of action, beginning with the concept of action-effect and reinforcement, i.e., feedback, is reminiscent of Leventhal's model of health behavior (see Chap. 2), in which the patient continuously evaluates the results of her 'therapeutic actions' and modifies the management of her treatment. Condition (3) of the volitionist model reminds us of a factor which appears in several models of adherence: The feeling of personal efficacy.

Finally, this new description of action requires a dynamic, rather than static, conception of adherence. We can see also that this power of will takes on additional importance in the context of a long-term health project; we shall emphasize this in the next part of this book.

References

Davidson D. Actions, reasons and causes. J Philos. 1963; 60: 685–700. Reprinted In: Essays on actions and events. Oxford: Clarendon Press; 2001.

Elster J. Emotions and economic theory. J Econ Lit. 1998;36:47–74.

Leventhal H, Falconer Lambert J, Dieffenbach M, Leventhal EA. From compliance to social-self-regulation: models of the compliance process. In: Blackwell B, editor. Treatment compliance and the therapeutic alliance. Amsterdam: Harwood Academic Publishers; 1997.

Livet P. Émotions et rationalité morale. P. U. F., Collection Sociologies; 2002.

Locke J. An essay concerning human understanding; 1690.

McNeil BJ, Pauker SG, Sox HC Jr, Tversky A. On the elicitation of preferences for alternative therapies. N Engl J Med. 1982;306:1259–62.

Narayan KM, Hoskin M, Kozak D, Kriska AM, Hanson RL, Pettitt DJ, Nagi DK, Bennett PH, Knowler WC. Randomized clinical trial of lifestyle intervention in Pima Indians: a pilot study. Diabet Med. 1998;15:66–72.

Proust J. La Nature de la volonté. Folio Essais: Gallimard; 2005.

Romains J. Knock. Folio: Gallimard; 1972.

Searle J. Intentionality, An essay in the philosophy of mind. Cambridge: Cambridge University Press; 1983.

Wolpert HA, Anderson BJ. Metabolic control matters: why is the message lost in the translation? The need for realistic goal-setting in diabetes care. Diabetes Care. 2004;24:1301–3.

Chapter 5
The Dynamics of Intentionality

Abstract In the Transtheoretical Model of Change described by Prochaska and DiClemente, the individual goes from a pre-contemplation stage, where there is not even a problem to be solved, to the stage of contemplation, where the person becomes conscious of the problem and considers resolving it. This is when the intention to adopt a health behavior is formed. The next step is to pass beyond preparation, to decide to adopt the behavior. The maintenance stage may then depend on the patient's resolution to treat herself. An analysis of the mental states of intention, decision and resolution, presented in this chapter shows the intervention of the force of habit, and of constructs such as self-control and willpower. One arrives to the idea that willpower shapes action: Certainly, desire and belief play a vital role in the production of action, but willpower is needed to bring the action cascade to fruition. For example, I join the gym because I want to lose weight and I believe exercise will lead to weight loss, but I actually embark upon an exercise regimen because I have willpower. I also employ willpower to persist in the regimen over time. The idea that willpower is a capacity that becomes tired with use helps us understand why individuals who chronically diet give up more easily when faced with temptation; why it is difficult to stay on a diet and at the same time quit smoking (Doctor, don't ask me to do everything at once).

5.1 Motivational Force

Up to now we have treated primary reasons, i.e. belief/desire pairs, as if they are bivalent—one either has a primary reason or one doesn't. But when we think about our desire for something, we not only think about that particular thing (the content of the attitude), but also about the strength of our desire. It's not just *that* we desire something; it's also a matter of *how much* we desire it. How can we characterize the strength of a desire, or what Alfred Mele calls motivational force? (Mele 1992, 11).

Mele starts with the assumption that since all intentional action is motivated action, it is in the strength of motivation that we find the strength of desire that

© Springer International Publishing Switzerland 2015
G. Reach, *The Mental Mechanisms of Patient Adherence to Long-Term Therapies*,
Philosophy and Medicine 118, DOI 10.1007/978-3-319-12265-6_5

plays a part in the primary reason of action. He notes that what motivates action A may not be the motivation of A, but rather the motivation of an action B that the person believes to be the result of action A: For instance, Gérard may be very motivated to avoid disappointing his daughter (B), and this motivation may motivate him to quit smoking (A). He does A to get to B.

This may contradict Davidson's conclusion that action B is not an action of the agent, because only actions such as A, i.e., primitive actions, are the true actions and that 'actions' such as B belong to the 'nature takes care of the rest' and for Davidson are part of the description of action A (Davidson 2001). This difficulty can be removed by assuming that the motivational strength of action B is the force of the agent's belief that accomplishing action A will bring about the event caused by B, whether B is 'performed' by agent or by nature.

Mele also points out that our evaluation of desire is more overtly concerned with its content than with its motivational force (Mele 1992, 37): For example, when we are in a restaurant, choosing between two favorite desserts, we make our choice based on what each dessert is, not by comparing the strength of our desires.

The motivational force, even if it is not apparent on first inspection (like the invisible pull of gravity that makes the apple fall down from the tree—all we see is the fall) does exist and can be quantified: The total motivational force of a desire is the sum of a positive motivational force and a negative motivational force (Mele 1992, 68) (we saw that, similarly, Nozick insists on the pro and contra, the positive and negative characteristics of the arguments that compose a 'reason'). The former involves all the attitudes (beliefs, fears, other desires) that contribute to the strength of a desire, and the latter all the ones that weaken it. This hidden layer of motivational push and pull is not always obscure: We may at times put our desires in abeyance and examine our deeper thoughts and feelings: What do I think I'm doing in this situation? Why am I getting so angry with this person? Is this what I truly want?

In Mele's model, our experienced desires are seen to be undergirded by a layer of complex, conflicting motivations. Consider, for example, the major role of fear of hypoglycemia in a patient's failure to adjust her insulin dose: Her desire to avoid severe hypoglycemia can weaken the strength of her desire to increase the insulin dose in the case of hyperglycemia. Hypoglycemia is so much more unpleasant than hyperglycemia that avoiding the former drives the (inadvertent) pursuit of the latter. This fear of hypoglycemia also explains the common observation that patients who don't usually increase their insulin doses in cases of hyperglycemia are prone to lower the dose after a hypoglycemia episode (Choleau et al. 2007). Only by examining the underlying motivations are such findings understandable.

For example, let's suppose that an individual wants to avoid the complications of diabetes, and she expresses this by adjusting her insulin dose (B); this in turn requires frequent measurements of her blood sugar (A). Suppose also that this action (A) seems to her to be the means to avoid severe hypoglycemia (C). Here, the desire to avoid severe hypoglycemia (C) appears as the way to reinforce the desire to measure her blood sugar frequently (A), which is motivated by the desire to avoid the complications of diabetes (it is part of the positive motivational force of A without being its cause).

Perhaps we have here the means to use a negative factor in a positive way: If you are afraid of hypoglycemia, measure your blood sugar often; this will help you adjust your insulin dose and avoid complications from diabetes. Avoiding long-term complications becomes as it were an offshoot of the patient's daily avoidance of acute hypoglycemia.

5.2 Self-control

Self-control is an ability which is well-understood in the psychology of daily life. It has, nevertheless, been confusing to understand: How can a person control herself? Doesn't this require a controller who is somehow distinct from the person? And who/what controls the actions of the controller? Like the problem of self-deception, self-control seems *logically* perplexing.

Alfred Mele describes self-control as an aptitude which enables a person to resist motivations that are opposed to the conclusions of 'all things considered' judgments, and it keeps these motivations from leading to an action that is contrary to these conclusions (Mele 1992, 54). Mele's self-control is fundamentally about restraint of lesser motivations; the task of how one determines which motivation should be dominant is not necessarily a simple one, however. In the previous section we talked about summing of positive and negative motivations as if this were a simple matter of arithmetic. In practice, it can be difficult to do this. For starters, it might be that the weight of each motivation is not clear; or that they sum to zero, or that the weights change over time (this last issue will be taken up later on). Indeed, these weights are hardly objective at all—as the term "weight" might mislead us to think—but the result of various individual factors. If I am too heavy, an attractive woman's criticism might spur me into exercise and dieting. But as my embarrassment wanes, so might the strength of my motivation. But it might also be that I see how well that woman's criticism has motivated me to do something I wasn't able to do before, and, as a way to keep up my flagging motivation, a photo of the woman is taped on my refrigerator as daily inspiration.

Sometimes we are insufficiently motivated to do what is "right"—and sometimes we are overly motivated to do what is "wrong." In this latter situation, we may need to use more explicit forms of self-restraint. When Ulysses' ship was to pass the Sirens singing their irresistible song, luring mariners onto deadly rocks, he commanded his crew to lash him to the mast so he couldn't steer his ship off course. Although we are rarely in such dire circumstances, we are certainly in less dramatic situations all the time. For instance, a dieter may keep sweets out of her house, knowing that in their presence she is likely to cave in and eat them.

Self-control may take the form of a character trait when at its behest the agent acquires the motivation, with regards to a given problem, to overcome motivations that undermine what she judges to be best for her (Mele 1992, 60). This does not mean that she will resist motivations contrary to her best interests in all situations (we saw that one can quit smoking, but not stay on a diet, and even more so when one no longer smokes). In addition, Mele reminds us that according to Aristotle, there is a continuum that goes from the people who have perfect self-control to those who never control themselves.

Techniques to initiate self-control have been explicitly used to facilitate the adoption of a health behavior such as smoking cessation or dieting (Fisher et al. 1982, 168–191; Rachlin and Green 1972). The concept of self-control used in these studies is very similar to the one developed by Mele. His concept is, in turn, akin to the one described by Skinner (1953, 236), who insists on the control of key elements of the environment but also on techniques to encourage the carrying out of unappealing tasks:

> In getting out of bed on a cold morning, the simple repetition of the command "Get up" may, surprisingly, lead to action.

Joan Miro, in an interview with Margit Rowell, described the effect of a visit to an exhibit of Jackson Pollock's works in the Facchetti gallery in Paris in 1952:

> Yes, certainly, this showed me a direction that I wanted to take but that up to then had remained at the stage of desire. When I saw this, I said to myself: "You can go there, go, you see, it's allowed" (Miro 1995).

And he painted the two paintings titled 'Fireworks', now at the Foundation Miro in Barcelona.

We can use this type of command every time we need to perform a therapeutic action (take a pill, refuse a cigarette, etc.). But we can also, once and for all, decide to have this attitude, i.e., *to get into the habit* of being adherent.

5.3 The Force of Habit

Is a habitual action the same as that action done only once? To the outside observer, a person performing a habitual action may look just the same as a person performing that action on a strictly as-needed basis: A person drinking a glass of milk in the morning—as he does every morning—seems the same as the person who drinks a glass of milk only because she ran out of orange juice.

Practical experience suggests that habitual action and one-time action differ. First, the habitual action is likely to be performed more skillfully; for this reason, it is more easily performed. Second, habitual action seems to require less thinking and effort. The person involved in a habitual action "just knows" what needs to be done; its automatic nature is one of habit's great benefits.

Davidson suggests that we even consider a series of repeated actions to be an action. He notes that

> the *sum* of all my droppings of saucers of mud is a particular event, one of whose parts (which was a dropping of a saucer of mud by me) occurred last night; another such part occurred tonight. [...] In these cases, we can talk of the same event *continuing*, perhaps after a pause (Davidson 2001, 183–184).

In Davidson's example, we might say, "I'm always dropping saucers"—by which we mean that we have a propensity to drop saucers, and do indeed drop them on occasion. If I should drop a saucer once, I regard this as an accident; but if I drop them over and over, I begin to see myself as having a single "saucer-dropping" habit.

5.3.1 Definition of Habit

A person can rise at four in the morning because today she wants to see the sunrise over the sea. Another person may wake at the same hour only on days when she travels out of town. And, finally, a third person may get up at 4 a.m. because she is a baker. In each case, the early-rising action requires the same motions and intention (expressed by setting the alarm), but it is only in the third case that we say the person is in the habit of getting up at four in the morning.

Similarly, if a patient took a pill yesterday, and again today, it is because she had the intention to do it yesterday and today (*repeated actions*) *or because, in general, she is in the habit of doing it.* She may, for example, be in the habit of taking her pill before breakfast. We can see that if this distinction is valid, the will of the patient who took the pill can express itself at two moments: (1) At the moment of performing the task (once or repeatedly), and (2) at the moment when the patient decided to get into the habit of doing it. If (2) has taken place, then (1) become obsolete, or is replaced by a (1'), where (1') can express itself by not breaking the habit. The idea here is that one does not need to decide to do something once it has become habitual—it has become second nature. But when the performance of the action requires an effort each time, it implies that one indeed makes the needed effort (this involves what we will see later: Volition and resolution). Similarly, in the opposite case, it is one thing to *never* take the pill, to be in the habit of not taking it, and it is another to not take the pill for, say, a day or a week.

This difference is important. We need to ask whether taking a pill everyday can truly become a habit. If this is the case, and if one can acquire this habit simply by deciding once and for all that this is how it is going to be, then we wouldn't need to develop repeated intentions in order to be adherent. Nonetheless, this would lead almost to a paradox: If no intention is necessary to take a pill, there is nothing to permit us to say that taking the pill is still an action. For, as you will recall, an intention is essential to an action. Without an intention, an action is simply a behavior, like putting one's coat on in wintertime; it requires neither reasoning nor the decision that one took at the end of autumn (it is beginning to get cold so from now on I will wear my coat instead of my jacket). The action becomes mindless and automatic.

Despite the fact that at least some patients will take their medication automatically and, as it were, mindlessly, still they must be capable of thinking about the action if called upon to do so. And indeed, patients are called upon to think about what they are doing: When they go to the pharmacy for a refill, or make an appointment to see the doctor for a new prescription, or have to report what medication they are on when seen by a new doctor.

That is why the act of taking a pill seldom becomes a full-blown 'habit' in the same way as the habit of tying our shoelaces after putting on our shoes. Taking medication is an interrupted habit, and is probably better understood as a series of repeated actions, even though some repeated actions do end up being truly habitual. One can, however, examine the problem from a different angle: The action was performed at the moment one decided to be adherent in the future. That is, the action was to get into the habit of taking care of oneself.

5.3.2 Mechanism of Habit

Let us first consider in more detail just what we mean by "habit."

A great deal of our daily lives is occupied by habits: Getting dressed each morning, eating breakfast, going to work, and so forth. Days are organized into weeks, the year is divided into seasons, we celebrate the same holidays year after year. In his 1834 work, "On Habit", Felix Ravaisson argued that habit is ingrained in the human mind, and that we cannot survive without habits (Ravaisson 1997). Though we may have a hand in which habits we cultivate, our habits also shape who we are.

Turning a therapeutic action into a habit may share some properties with what Mele calls self-control.

Self-control: Take Mele's following example of self-control: Ian is sitting in front of the TV watching a golf tournament, but he needs to repaint the shed in order to surprise his wife. This looming project annoys him, however, and he continues to watch TV instead. So Ian says aloud to himself: "Get off your butt, Ian, and paint that shed!" He then gets up and goes to the backyard to repaint the shed.

Habit: Consider again our prototypical scenario—a patient must take (or not take) an aspirin prescribed for prevention of cardiovascular complications. Our patient has a desire to preserve her health, and the motivational force of the desire is strong, but she is also skeptical that aspirin will do what her doctor has said it can. Moreover, she doesn't wish to spend any time on this boring activity, especially if it shall be ineffective. On top of that, she knows that she is the sort of person who, if she makes a commitment to take the medication, will be anxious and preoccupied when and if she misses a dose. Is taking the aspirin really worth it? But this patient is adherent; she gets into the habit of taking the aspirin before breakfast, and she keeps the bottle of pills next to the coffeemaker. Thanks to this habit, she takes the aspirin, *even if her desire not to do it is stronger.* It is easier to do something boring and repetitive when one is in the habit of doing it.[1]

---

[1] This copies Mele's demonstration, replacing the notion of self-control (C in his demonstration) with the notion of habit (H). Due to habit (H), the patient performs the continent action (A) of taking the aspirin, even if the desire not to do it (NA) is stronger: let's suppose that the motivational force of the desire to take the aspirin (A) is weaker than that to not take it (NA), which we can describe by fm(A), for example force 4 < fm(NA), for example force 12. We can suppose that the positive motivational base of the habit of taking the aspirin (H) is the same as the desire to do (A), DA. But, and this is the key to Mele's demonstration, since the positive motivational base of the desire does not explain all of its motivational strength, it does not follow that the motivational force of the habit of taking the aspirin fm(H) is less than the motivation force of the desire to not take the aspirin fm(NA). It can also be much stronger (for instance force 20) and it can be easier to appeal to habit than to give into the temptation of missing a dose of the medication.

Habit not only makes the task easier, but it also makes us less inclined to break the habit—because the act of breaking becomes more effortful than just doing the original task.

Habitual actions are easier to perform, as was noted by Democritus (Fragment B241)—Continuous labor becomes lighter by custom—, and more recently by Hume (2008, 302):

> When the soul applies itself to the performance of any action, or the conception of any object, to which it is not accustomed, there is a certain unpliableness in the faculties, and a difficulty of the spirit's moving in the new direction. As this difficulty excites the spirits, it is the source of wonder, surprise, and of all the emotions, which arise from novelty; and is in itself very agreeable, like every thing, which inlivens the mind to a moderate degree. But though surprise be agreeable in itself, yet as it puts the spirits in agitation, it not only augments our agreeable affections, but also our painful, according to the foregoing principle, that every emotion, which precedes or attends a passion, is easily converted into it. Hence very thing, that is new, is most affecting, and gives us either more pleasure or pain, than what, strictly speaking, naturally belongs to it. When it often returns upon us, the novelty wears off; the passions subside; the hurry of the spirits is over; and we survey the objects with greater tranquility.

Note the analogy with the role of novelty for the birth of emotions proposed by Livet (2002).

We shall now see that habit has a central role to play in treatment adherence.

5.3.3 Advantages of Habit

The resort to habit, for example when taking a pill, can have many advantages. First, it helps one avoid all the unpleasant thoughts related to one's illness each time one takes it. So it can be an effective coping technique when facing a chronic disease. Second, we just saw that it is easier to perform repetitive and boring tasks by appealing to habit. And finally, habit can be a powerful alternative to memory. A first approach is the use of reminders such as pillboxes and written notes. The use of habit could be another solution, perhaps even a necessity: The patient needs to remember to use the pillbox or to look at the instructions, which is simply the replacement of one act of memory by another. Again, it is easier to get into habit of doing it (Reach 2005).

5.3.3.1 Patient Adherence: Shuttling Between Habit and Deliberation

Habit has its downsides as well. Émile Durkheim noted both the benefits and problems associated with habit:

> Habits tend to excite active phenomena. Habits tend to diminish the intensity of passive phenomena. When a psychological phenomenon is active, the habit excites it, making it still more active and allowing the phenomenon to be more easily reproduced. But if the phenomenon in question is passive, habit weakens it, even to the point that it becomes imperceptible (Durkheim 2004, 152).

For example, a habit such as "I will drink a diet Coke every evening instead of a beer" is more likely to self-reinforce, whereas a habit such as "I won't drink beer in the evening" is more likely to peter out.

The "mindlessness" of habits can be perilous as well. If a habitual behavior calls for some deliberation—say, checking that one has grabbed the correct medication bottle from the cabinet—one might not notice that one holds the wrong bottle. Or, one might mindlessly take a pill from the correct bottle, but not notice that the pharmacist accidentally dispensed the wrong medication. Thus habit facilitates the performance of repetitive tasks, but there is also the risk of a routine that can be dangerous for the tasks that require some care.

In the treatment of diabetes, the routine injection of insulin creates the danger of forgetting that it is necessary to adjust one's insulin dose: Leventhal argues that neither habit nor deliberation ever do well as strategies on their own, but must depend upon one another. Too much habit, and needless accidents occur; too much deliberation, and nothing gets done:

> A wide range of compliance producing procedures may be involved in a treatment regimen. Our hierarchical model of the processing system suggests that some procedures must address the patients' conscious conceptual planning, while others must address automatic reactions. For example, conscious planning, such as list and appointment making, is necessary to obtain and fill prescriptions, while automatic, environmentally elicited responding is best for consistent pill taking (e.g., keeping medication with one's breakfast cereal.) (Leventhal 1997, 25).

This has important implications for patient nonadherence in the treatment of chronic illness. Habit, counterpoised with deliberate action, is one way that open-ended treatment is accomplished. What has been called the ritualization of the therapy is not a mere crutch for the forgetful, but is vital for successful adherence over the long run. In the first part of the book we have mentioned that ironically, adherence to a treatment with a pill before each meal (the concept of one meal, one pill) can be good, contrary to the idea that it is better to encourage medication requiring only one daily dose.

As Mele has noted, self-control relies not only the ability to decide to activate it, but also the skill to find ways actually implement it (Ulysses having himself tied to the ship's mast). In medical care, self-control might consist of finding ways to reinforce habit: Putting one's pill bottle next to the coffeemaker, or rewarding oneself with a massage after going to the gym to exercise once a week. As Janine Pierret notes concerning treatment of AIDS,

> the enforcement of treatment is going to be accomplished by progressively adopting a routine, most often in daily life. Taking the medication will be attached to particular moments of the day and in sync with different activities: medication taken immediately after getting up, when coming home from work, at dinner. This is why forgetting and skipping doses are most often connected to changes in the rhythm of work and leisure and are almost never the result of the medication itself (Pierret et al. 2001).

5.3.4 Training Through Habit

Yet habit does not settle in through mere decision, nor does it become habit immediately: It takes time to grow and mature. And, like many things which grow and mature, habit can also wither and die. Getting a habit started and then sustaining it is a skill in itself, above and beyond the particulars of some habitual activity (perhaps this is why the war on smoking is only won after a certain period of abstinence, as is suggested by Prochaska's Transtheoretical theory of change). Nurturance of habit—once seen as an important facet of the sober, mature life—is now often overlooked in modern society, or seen as a negative: Bad habits, a drug habit, etc.

There are certainly a good number of habits that we get into without ever having *explicitly decided* to adopt them; we "fall into" them simply through the repetition of actions that at first were intentional. Clearly we are back to the mechanisms pertaining to behaviorism: If the performance of an action resulted in pleasure or displeasure, the mechanisms of conditioning through positive or negative reinforcement come into play.

The effect of Antabuse engages the same mechanism: When taken at the same time as alcohol, it induces vomiting. More recently, a medication inhibiting the absorption of fats was introduced in the treatment of obesity. The patient must avoid eating fatty food in order to avoid an extremely unpleasant greasy diarrhea. During the treatment the patient learns to do this and one hopes that by the end of the treatment she will have gotten into the habit of doing it.

As "creatures of habit", people sometimes do not even realize they have a particular habit until it is interrupted: For example, I listened to a radio program on my drive to work simply because the radio was set to that station, and I didn't care what was playing so long as there was something to keep me alert. But when the program host moved to another time slot, I was bereft. In short, we sometimes are like Skinner's lab rats, learning patterns of behavior through mindless repetition and reinforcement, and even not aware that we are being reinforced.

Damasio's hypothesis of 'somatic markers', which he has proposed as an explanation of how emotions help to shape our choices, is not very far from this behaviorist mechanism of decision-making:

> When the choice of option X, which leads to bad outcome Y, is followed by punishment and thus painful body states, the somatic marker system acquires a hidden, dispositional representation of this experience-driven, noninherited, arbitrary connection. Reexposure of the organism to option X, or thoughts about outcome Y, will now have the power to reenact the painful body state and thus serve as an automated reminder of bad consequences to come (Damasio 1994, 180).

When a therapeutic task is difficult and calls for deliberation, having repeatedly accomplished it before diminishes the need for deliberation each and every time: That is, *practice makes perfect*. Let's take for example the task of adjusting one's insulin dose, which at first glance seems to call for the use of complicated rules. One wonders if at a certain point the patient who has done it repeatedly does not end up adjusting her doses without analytically using the rules. The acquisition of mastery in performing any skill is often imagined to be a progression from concrete experiences to ever more abstract, general concepts. Hubert Dreyfus argues that the development of expertise actually follows just the opposite course (Dreyfus 1992, 352–73; Dreyfus and Dreyfus 2004). The novice starts out by applying the abstract rules that she has been taught without paying much attention to the context; as she develops, she enriches the old rules with new rules that she discovers by application to particular cases, and which she experiences in context (one is tempted to say in the holistic context of her mind). She then tries out these rules by introducing a personal strategy for which she will feel responsible, experiencing a personal satisfaction when she witnesses its success and displeasure at its failure. Satisfaction and displeasure doubtlessly play a role in the selection, not of the rules, but of the situations where they will be applied. And, finally, she becomes capable of treating the particular cases without using the theoretical rules independent of context when she recognizes an analogy with previously experienced situations for which she had found satisfying solutions.

In our example, the patient becomes capable of choosing the adjustment to the insulin dose, perhaps not quite like tying her shoelaces, but without having to accomplish the fastidious analytical work of selecting the rules. And if we ask her how she reasoned, which rules she used, she would not be able to explain it because, as Dreyfus puts it,

> the expert simply does not use rules! If one asks the expert about rules, one forces him to regress to the beginner level and to list the rules that he still remembers, but that he no longer uses.

An identical path is followed in the acquisition of physical skills, and here we find once again that the Background is pertinent. John Searle describes the training of a skier: At the beginning, he knows that he must put his weight on the downhill ski. But there comes a time when he no longer uses this rule, and if he did, he would be impeded in his movements. For Searle,

> 'practice makes perfect' not because practice results in a perfect memorization of the rules, but because repeated practice enables the body to take over and the rules to recede into the Background (Searle 1983, 150).

This is similar to what Joelle Proust says when she stresses that,

> because of the accumulated experience of acting, real consciousness is no longer essential for the actions with the most familiar consequences. The sequence volition-transformation, once steeped in the habits of acting, frees the attention for the more demanding tasks. This is why attention gradually moves from the level of motor realization to the level of attaining goals further removed from the actual execution (Proust 2005, 144).

It is obvious that such mechanisms play a role in patients' decisions: When one advises a patient concerning the modification of insulin doses using classic rules, often the patient answers, "yes, but if I do this, I run the risk of hypoglycemia", basing her answer not on a theoretical rule, but on previous experience; it is a foolish doctor indeed who dismisses this out of hand. As Joseph Nuttin notes, reasoning by analogy is not mere conditioning, but an "instrumental use: Successful action is perceived as an efficient method to reach a goal". The ability to recognize the distinct identities of problems that are formally identical could differ among individuals and this difference may play a role in the ability to develop this type of reasoning-based action (Nuttin 1980, 72, 73).

5.3.5 Adherence to Long-Term Therapies: A Habit of Action

Certainly, then, a vital step is to *begin* taking care of oneself. And it is therefore at the beginning that the patient must be offered special support. Thus, I propose that one can foster adherence by deciding to acquire the habits of adherence—or, for those not quite so ready, to be willing to (at some point) decide to be adherent.

The patient is basically saying, "I'm not ready to make a change now, but I am willing to be open to the idea of change in the future." Some patients might, of course, use this as a way to avoid change indefinitely; but for many patients, it keeps them psychologically engaged in the change process, and often brings out the issues which make them ambivalent about change in the first place.[2]

One can replace the need to make choices for isolated actions with the practice of repeated actions performed thanks to the force of habit. One then chooses to get into the habit of taking care of oneself, and it is at that moment that one has "considered all things": One decides to weigh the two sides of the conflict between the desire that makes one take the pill and the one that makes one not take it. This attitude will protect us in the future from the temptations leading to nonadherence. In short, we make a "meta-commitment" to self-care, and from this naturally flows various self-care actions (Fig. 5.1).

The daily choice and the primitive choice of being adherent in the future by getting into the habit of performing the therapeutic act, making the daily choice useless. Repeated actions can favor the formation of habit, which is shown by the double arrow; it also makes the need for reasoning less pressing. But it also has the advantage, by leading to the phenomenon of habituation described by Pierre Livet, to limit the force of emotions (fear) that were present during the announcement of the illness and to combat anxiety stemming from the impossibility of following the change of preferences represented by the primitive choice. At the right of the figure, a reminder that according to Davidson the results of the action do not depend on the agent but on nature "that takes care of the rest".

In this model of changing priorities, we again find Livet's process of revision by which emotion arises from perceived differences between how we imagine things to be, and how they really are. In the case of habit formation, the initial decision to

---

[2] I'm grateful to John Meyers for this remark.

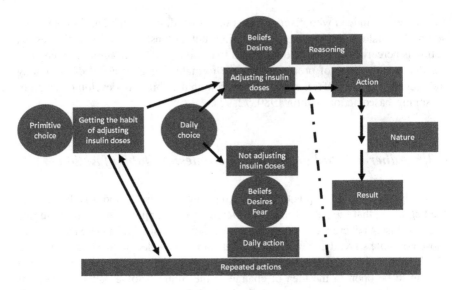

Fig. 5.1 The two moments of choice, in the example of patient adherence and nonadherence, to adjust the insulin dose or not. Reproduced from Reach (2000). Copyright © 2000 Elsevier Masson SAS. All rights reserved

take care of oneself introduces just such a disequilibrium: Previously, one thought all was well, or at least well enough; but in the new frame of self-care, one sees work to be done. Livet's model should not be understood as implying that cognition precedes emotion; but that our beliefs (cognitions) and emotions are tightly coupled—Livet himself notes that changes in beliefs may even be unconscious:

> most of the time, we don't think about revising our beliefs, [until] one day we notice that we no longer feel the emotion, or only in a weakened form, and that our preferences have changed. The revision can be largely unconscious (Livet 2002, 75).

The revision of preferences allows one, over time and through the phenomenon of habituation, to channel the force of emotions and avoid the anxiety stemming from a state of perpetual indecision. A patient may not know exactly where her determination to take care of herself stands; to put this in terms of Prochaska's helices of change, for example: She can progress imperceptibly on the path towards adherence; a conscious change in beliefs is neither necessary nor sufficient.

Clinical experience shows that this revision is, nevertheless, often a conscious process. It can be an authentic rational choice reached at the conclusion of true deliberation. Many former smokers can give the date of their last cigarette; recovering alcoholics routinely report the date of their last drink. Other aspects of adherence seem to be similarly conscious: Three years ago, having "considered all things", I finally decided to be reasonable and stop arguing and opposing those who sought to counsel me; I hadn't even gotten to the specifics of my unreasonableness, but I had decided to lower my guard and stop fighting.

We shall soon see that George Ainslie suggests a similar solution to fight weakness of will (Ainslie 1999): Unify one's choices into a single, momentous decision; he calls it 'bundling': It is the creation of personal rules. Good habits are the armor of the weak-willed: They sometimes allow the person who has firmly decided to stop smoking to indulge in a good cigar after dinner without putting the future of the decision at risk, creating exceptions that confirm one's commitment to the rule.

One may now understand the wisdom of taking a break once in a while, as is advised by the Stoic Seneca, for whom "there is a great difference between simple living and slovenly living", the latter leading to the risk of falling back into bad habits. Ainslie also shows that the strict use of personal rules is not without danger if we do not once in a while 'take a break' (Ainslie 1999). Exceptionally rigid adherence can, at least sometimes, lead to disastrous failures.

5.4   Intention, Decision, Resolution, and Willpower

Context is all-important. Facts are facts unchangingly, but meanings depend vitally on context—and it is meanings which drives behavior, not facts. We are dealing with much more than just *performing the action* of 'taking the pill': The patient must have the intention to take care of herself. In some way, the successful patient must engage in the action because it has a meaning for her. Meaning is not a prerequisite to action, and it needs not be static over the course of treatment, and it is not even always clearly articulated—but it is always present in successful therapy.

In Prochaska's model (Prochaska and DiClemente 1983), the individual goes from the pre-contemplation stage, where there is not even a problem to be solved, to the stage of contemplation, where the person becomes conscious of the problem and considers resolving it. This is when the intention to adopt a health behavior is formed. The next step is to pass beyond preparation, to decide to adopt the behavior. The maintenance stage may then depend on the patient's resolve to treat herself (Fig. 5.2). An analysis of the mental states of intention, decision and resolution will allow us to better understand this model.

5.4.1   The Notions of Intention and Decision

In the *Nature de la volonté* (*Nature of the Will*), Joelle Proust shows how Searle's introduction of intention to the philosophical theories of action was remarkably innovative (Proust 2005, 82). Prior to Davidson, the driving force behind action was seen as the conjunction of a desire and a belief. Searle suggests that there is a mental state, the intention, which cannot be reduced to beliefs and desires and which

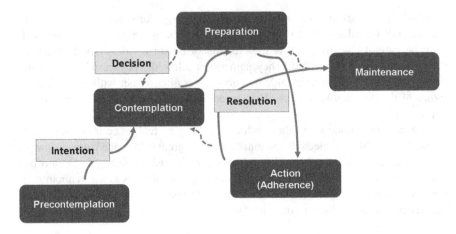

Fig. 5.2 Intention, decision in the example of patient adherence and Prochaska's model

is the true impetus to action. Only if a behavior is caused by an intention can it be qualified as an action. Interestingly, this was Sartre's definition of action as well.

We have defined *intentional* states as mental states that have a content: Intention and decision are particular types of intentional states (Searle 1983, 79–111). And just as belief has a condition of satisfaction, so does my decision. For instance, if I decide to take this pill, the decision is satisfied—completed, brought to a close—when I actually take the pill.

The situation is usually more complicated for intention: I may have the intention of exercising, but I may never realize this intention. I can remain at the stage of contemplation indefinitely. Generally speaking, decision usually leads to an action; but intention has the potential to linger, sometimes for very long times indeed. It is this deferred intention which characterizes the contemplation stage. On the other hand, when one enters Prochaska's preparation stage, the transition to 'action' generally occurs usually within one month.

Michael Bratman points out another difference between intention and decision: Intention is linked to a belief, and like belief, intention is independent of context—it is irrational to give up believing that water is wet simply because I have moved from Majorca to Madrid. It is this independence from context that allows intention—and belief—to persist over time. On the contrary, a decision to do something is connected not to a belief, but to acceptance: In this case there was a moment, with an exact date, when I decided that the content of the belief was true. As opposed to belief, acceptance depends on context and is the result of a voluntary action. Consequently, decision, unlike intention, depends on context (Engel 2000; Bratman 1999, 33). Now we can understand the role that external factors play in the *Health Belief Model* (see Fig. 2.1): It may be a change in context that leads to a patient's decision to take care of herself.

5.5 The Dynamics of Intentionality

For Davidson, to form the intention to perform an action at a given moment means, at that moment,

> to hold that it is desirable to perform an action of a certain sort in the light of what one believes is and will be the case (Davidson 2001, 100).

Nonetheless, as Michael Bratman notes Davidson gives little information concerning specifically 'future' intentions, the ones that involve actions that will not be performed immediately (Bratman 1999). And yet these 'future' intentions play a key role in adjusting means to ends, making future and past projects coherent. Searle makes a similar distinction: Between 'prior' intention, for which the condition of satisfaction is the entire action, and the intention 'in action', where the condition of satisfaction is the precise bodily movement that I perform at this moment. In the case of patient adherence, then, intending to take care of oneself is a prior intention while intending to take a pill is an intention in action. Here is Searle:

> And thus the prior intention causes the intention in action. By transitivity of Intentional causation, the prior intention represents and causes the entire action, but the intention in action causes only the bodily movement (Searle 1983, 95).

Elisabeth Pacherie suggested, with greater precision, that when considering a long-term project there are three types of intentions: Those oriented towards the future (F-intentions), those dealing the present (P-intentions) and motor intentions (M-intentions). These intentions differ in their role regarding the realization of the project, the type of content that is associated with them, temporal constraints to which they are subject, and their dynamics. They intervene successively, and as each one inherits its goal from the preceding intention, their unity in the context of a single project is guaranteed (Pacherie 2003).

5.5.1 To Take Care of Oneself Day After Day: An Interpretation Within the Framework of a Theory of Intentionality

We have defined patient adherence in the case of chronic illness as "the acceptance to perform repeatedly a series of *actions* prescribed with the objective of long-term health". Let us now consider the different phases of the health project of adhering to treatment for a chronic illness, and distinguish the general long-term context from the more up-close events of care. In particular, we have noted the often 'mechanical' character of actions such as taking one's medication each day. Are these mechanical, automatic actions truly actions, or do they differ in some way? Elisabeth Pacherie's model suggests that these automatic actions are 'minimal' actions. Minimal actions are not preceded by any form of practical deliberation, conscious or otherwise.

This leads us once again to stress the role of habit and routine in patient adherence. The ritualization of medication-taking can explain the adherence to the pills taken before each meal: *One meal, one pill.*

In Searle's scheme of intentional actions, the intention in action (to take a pill) has as its condition of satisfaction my actually taking the pill; it is *this* intention in action that must cause *this event* (someone could put the pill in my mouth and make me swallow it, but in this case taking the pill would not be my action). This intention in action is generally preceded by a prior intention (to take care of myself) where the condition of satisfaction is an *action*, to take the pill. The action consists of two components: The *event* of taking the pill and the *intention in action*. *This prior intention* causes the *action*. Thus, it does not satisfy the prior intention if, today, I take the pill for a completely different reason, such as to finish this bottle of pills because I don't want to see it anymore.

Even though prior intention can lead to a specific action, it is not true that all specific actions are preceded by a prior intention. Searle demonstrates this by a comparison of visual perception and memory:

> neither the memory nor the prior intention is essential to the visual perception or the intentional action respectively. I can see a lot of things that I have no memory of seeing and I can perform a lot of intentional actions without any prior intention to perform those actions (Searle 1983, 96).

5.5.2 Back to the Mechanism of Habit

Habit, then, removes the need for the intervention of a prior intention. An effective habit can exist by itself, without having to appeal to a prior intention. Because forming prior intentions require some concentration and deliberation, habits are a sensible way to lessen, and perhaps streamline, our cognitive load (Fig. 5.3).

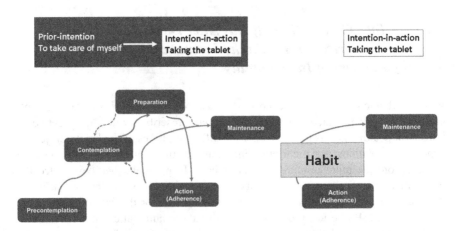

Fig. 5.3 The mechanism of a habitual action (adapted from Reach 2005). Originally published in (Reach 2005). Modified with kind permission of © John Wiley and Sons 2004

At the beginning of the treatment (left panel), taking a pill is preceded by a prior intention. Habit removes the intervention of prior intention (right panel). The initial stages of Prochaska's cycle have fallen by the wayside once a habit is formed and active. Perhaps the disappearance of these preliminary stages is a safeguard against retrogression? The past seems to have disappeared from the patient's mind when habit is functioning well. Despite the well-known risks of forgetting one's past, such forgetting may have benefits as well.

5.5.3 Resolution and Willpower

Regarding habit as a passage to a state of routine in which prior intentions disappear nonetheless poses a problem: It seems to strip the action of its intentionality. We have been arguing that adherent actions are intentional—and yet one of the best ways of achieving adherence seems to be non-intentional! And yet we have also argued that patient adherence requires a continual give and take between habit and deliberation. The successfully adherent patient must eschew the extremes of mindless, Skinnerian habit and paralyzing, time-consuming deliberation.

Is habit the only mechanism for persistence in a health project? Richard Holton suggests that an action is not determined by a desire and a belief alone or even by a belief, a desire and an intention alone. He suggests, rather, that there is also a person's *willpower* (Holton 2003). A person has the "strength" to stick to her intention, according to Holton, by virtue of her willpower. It counteracts the temptation to "give in". Let's say that willpower is a distinct phenomenon which determines how likely a person is to transition from prior intention to intention in action, and from intention in action to the act itself. Since our intentions not infrequently come to naught, it seems sensible to wonder if another factor may intervene to alter the cascade from intention to act.

The return of the notion of willpower in a theory of action cannot but remind us of Joelle Proust's theory of volition. Proust uses John Locke's classic definition of the will (1689):

> a *Power* to begin or forbear, continue or end several actions of our minds, and motions of our Bodies (*On Human Understanding*, II. XXI 5. 7–11).

She notes that this definition

> identifies a series of capacities (...) that have three types of application: they must allow one to *begin* an action (which might involve suspending action until a favorable time for its execution), to *forbear or carry it out to its conclusion* once the action is begun (by eventually correcting what needs be), and finally to *end the action* if and only if the goal is considered to have been reached. Volition is, in this analysis, the mental act that allows one 'to begin an action' (...) it is the conscious executive act that puts intentional states such as desires and intentions to work (Proust 2005, 124).

But volition does not intervene only in isolated actions, but also in actions that are part of an entire project. Proust gives the example of crossing a town square:

> A single volition is necessary, beginning with the first step and ending with the last, the intermediary stages being the proof that let one measure the shrinking of the distance to one's goal;

or, the project of writing a novel or a philosophical treatise:

> where we can obviously show that the agent cannot perform this action in one sitting, requiring the formation of distinct volitions, as is proven by the subjective feeling of having to renew one's efforts (Proust 2005, 148).

Let me propose that Holton's willpower is nicely suited to the job of forming distinct volitions in order to complete a complex project spread out over time (in our case, the health project). Holton believes that some of our intentions can be qualified as resolutions: Intentions capable of overcoming inclinations detrimental to one's long-term project. If a new desire turns out to be stronger than the usual intention, and if the agent changes her mind, we will simply call her fickle; if the new desire takes control of the resolution, we will say that she has demonstrated a weakness of will. But certain individuals are capable of carrying out the goals they have fixed for themselves. They succeed because they have a special capacity that allows them to mentally repeat what they have committed to—because one must do it to persist—without reconsidering the decision.

This is precisely what willpower does. Holton suggests that this ability, like any muscle, has limited capacity, tires little by little, and varies in its efficacy with context (thus former alcoholics have a higher chance of falling off the wagon if they are tired, depressed or anxious).[3]

For Holton this gives us reason to think that willpower exists by itself and intervenes in all of a subject's projects. Like muscular force, one can develop willpower by exercising it often (as one becomes virtuous by practicing virtue).

Willpower, then, shapes action. Certainly, desire and belief still play a vital role in the production of action, but willpower is needed to bring the action cascade to fruition. For example, I join the gym because I want to lose weight and I believe exercise will lead to weight loss, but I actually embark upon an exercise regimen because I have willpower. I also employ willpower to persist in the regimen over time.

So far, this idea of willpower is commonsensical. But next, we will explore a more controversial notion about willpower, namely, that it can be depleted. In general, we tend to imagine willpower as a moral quality, which some people have plenty of, and others less. Yet on closer inspection, we also recognize that individual people can run out of willpower: When exhausted, stressed or harried, our ability to exert willpower is diminished (one might even see this as adaptive, as willful action may be less successful if done when in a depleted state). Willpower seems to get used up, and time is needed to regenerate it.

---

[3] In Alcoholics Anonymous, the acronym HALT—Hungry, Angry, Lonely, Tired—is used to remind alcoholics of these dangerous contexts, and that they should "halt" when any of them arise, and fix the situation. I am grateful to John Meyers for this remark.

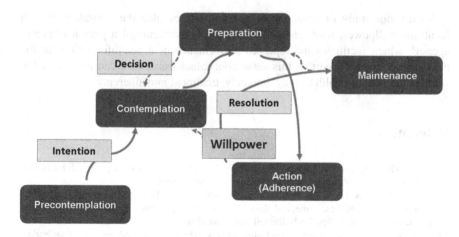

Fig. 5.4 Willpower, another way to keep one's resolutions

The idea that willpower is a capacity that becomes tired with use helps us to understand why individuals who chronically diet give up more easily when faced with temptation; why it is difficult to stay on a diet and at the same time quit smoking (Doctor, don't ask me to do everything at once). This hypothesis also allows us to understand the effect of powerful emotions, of constant decision-making, or of alcohol, on one's commitments. All of these situations diminish one's willpower (Baumeister and Vohs 2003, 201–216). "Willpower", according to a recent hypothesis (fascinating for a diabetologist!) may be more than a metaphor: Willpower may rely on glucose as a limited source of energy (Gailliot et al. 2007), and some studies suggest that acts of self-control like emotion regulation deplete blood glucose levels. However, this occurs only in poor emotion regulators (Niven et al. 2013).

We can also gird up our willpower, at least temporarily: When overweight patients are asked what they expect from a gastric ring that will force them, by mechanically constricting their stomach to decrease the size of their meals, they often answer that a gastric ring will be a somewhat forced, mechanical aid to their weakness of will. A patient said before bariatric surgery: "I knew that nothing was going to change unless something stopped me" (Ogden et al. 2006). We bolster our will by engaging in self-talk, posting reminders and affirmations on the bathroom mirror, and empowering friends and family to remind us to stay on track.

We can now understand that Prochaska's maintenance phase can be strengthened by a capacity other than habit (Fig. 5.4). This is fortunate since, as we have seen, habits may slip into routines and thence into disuse. Willpower is an alternate and indeed preferable mental resource to ensure adherence to long-term therapies.

It may be that the use of habit is more effective for simple, concrete actions such as taking a pill, while willpower is perhaps more useful for maintaining more abstract courses of behavior, such as staying on a diet, exercising, and not smoking.

So far, our study of treatment adherence has revealed the centrality of both habit and willpower; habit and willpower are components of a person's "health agency" which facilitates the repeated performance of a healthful action. In the next chapter we will see that this same conceptual framework can illuminate the vexing dark side of health agency—namely, treatment nonadherence.

References

Ainslie G. The dangers of willpower. In: Elster J, Skog O-J, editors. Getting hooked, rationality and addiction. Cambridge: Cambridge University Press; 1999.

Baumeister R, Vohs K. Willpower, choice and self-control. In: Loeweinstein G, Read D, Baumeister RF, editors. Time and decision, economic and psychological perspectives on intertemporal choice. New York: Russell Sage Foundation; 2003.

Bratman ME. Faces of intention, selected essays on intention and agency. Cambridge: Cambridge Studies in Philosophy; 1999.

Choleau C, Albisser AM, Bar-Hen A, Bihan H, Campinos C, Gherbi Z, Jomaa R, Aich A, Cohen R, Reach G. A novel method for assessing insulin dose adjustments by diabetic patients. J Diabetes Sci Technol. 2007;1:3–7.

Damasio AR. Descartes' error. Emotions, reason and the human brain. New York: G. P. Putnam's Sons; 1994.

Davidson D. Actions, reasons and causes. J Philos. 1963;60:685–700. Reprinted In: Essays on actions and events. Oxford: Clarendon Press; 2001.

Dreyfus SE, Dreyfus HL. A five-stage model of the mental actvities involved in directed skill acquisition; 1980. Reprinted In: Bull Sci Technol Soc. 2004;24:177–181.

Dreyfus HL. La portée philosophique du connexionnisme. In: Andler D, editor. Introduction aux sciences cognitives. Paris: Gallimard; 1992.

Durkheim E. Philosophy lectures, notes from the Lycée de Sens, 1883–1884 (trans: Gross N, Jones RA). Cambridge: Cambridge University Press; 2004.

Engel P, editor. Believing and accepting (Philosophical studies series). Dordrecht: Kluwer Academic Publishers; 2000.

Fisher EB, Levenkrohn JC, Lowe MR, Loro AD, Green L. Self-initiated self-control in risk reduction. In: Stuart RB, editor. Adherence, compliance and generalization in behavioral medicine. New York: Brunner Mazel; 1982.

Gailliot MT, Baumeister RF, DeWall CN, Maner JK, Plant EA, Tice DM, Brewer LE, Schmeichel BJ. Self-control relies on glucose as a limited energy source: willpower is more than a metaphor. J Pers Soc Psychol. 2007;92:325–36.

Holton R. How is strength of will possible? In: Stroud S, Tappolet C, editors. Weakness of will and practical irrationality. Oxford: Clarendon Press; 2003.

Hume D. Treatise on human nature, section V. Sioux Falls: NuVision Publications; 2008.

Leventhal H, Falconer Lambert J, Dieffenbach M, Leventhal EA. From compliance to social-self-regulation: models of the compliance process. In: Blackwell B, editor. Treatment compliance and the therapeutic alliance. Amsterdam: Harwood Academic Publishers; 1997.

Livet P. Émotions et rationalité morale. P.U.F., Collection Sociologies; 2002.

Locke J. On human understanding; 1689.

Mele AR. Irrationality, an essay on akrasia, self-deception and self-control. Oxford: Oxford University Press; 1992.

Miro J. In: Lelong D, editor. Écrits et entretiens; 1995.

Niven K, Totterdell P, Miles E, Webb TL, Sheeran P. Achieving the same for less: improving mood depletes blood glucose for people with poor (but not good) emotion control. Cogn Emot. 2013;27:133–40.

Nuttin J. Théorie de la motivation humaine. P.U.F., Collection Psychologie d'Aujourd'hui; 1980.

Ogden J, Clementi C, Aylwin S. The impact of obesity surgery and the paradox of control: a qualitative study. Psychol Health. 2006;21:273–93.

Pacherie E. La dynamique des intentions. Dialogue. 2003;42:447–80.

Pierret J, Study Group APROCO. Une approche dynamique du traitement chez des personnes infectées par le VIH: la notion d'intégration. In: L'Observance aux traitements contre le VIH/SIDA, Agence Nationale de Recherche sur le SIDA; 2001.

Prochaska JO, DiClemente CC. Stages and processes of self-change of smoking: Toward an integrative model of change. J Consult Clin Psychol. 1983;51:390–5.

Proust J. La Nature de la volonté. Folio Essais, Gallimard; 2005.

Rachlin H, Green L. Commitment, choice and self-control. J Exp Anal Behav. 1972;17:15–22.

Ravaisson F. De l'Habitude. Collection Rivages poche, Petite Bibliothèque, Foreword by Frédéric de Towarnicki, Payot, Paris; 1997.

Reach G. Application de la théorie causale de l'action à l'analyse de la non-observance thérapeutique. Presse Médicale. 2000;29:1939–46.

Reach G. The role of habit in therapeutic adherence. Diabet Med. 2005;22:415–20.

Searle J. Intentionality, an essay in the philosophy of mind. Cambridge: Cambridge University Press; 1983.

Skinner BF. Science and human behavior. New York: MacMillan; 1953.

Chapter 6
Medical Irrationality

Abstract Nonadherent patients are often conscious of their behavior but do not understand it: When asked why they did not stick to their diet, obese persons often do not try to hide their nonadherence but add with a sigh—"I know, I should, but I can't help myself." This perplexing human behavior has been described under the name of akrasia (literally lack of strength). Other philosophers have used the term incontinence, or weakness of will. In this chapter, I propose to describe nonadherence as a case of incontinent action. The definition of an incontinent action is that the agent intentionally performs an action that she herself does not consider to be the best—not an action that is considered bad by another. In short, the classic view of nonadherence was about disagreement between doctor and patient, whereas considering nonadherence as a case of akrasia places the disagreement between the patient and herself. Davidson suggested that there is a principle of continence: When one abides by this principle, one commits to using all the available information before acting (this is the 'all things considered'). Akrasia is the consequence of a failure of this principle. To explain how the exile of the principle of continence is possible, Davidson proposed the hypothesis of a divided mind. This partitioned mind hypothesis is applicable to clinical experience: For instance, there are pipe smokers who take the warning labels out of the tobacco boxes so that they do not have to 'think' about it, or perhaps because the labels confront their irrationality.

We have sought to better understand the paradoxical quality of patient nonadherence, and our seeking has led us to the field of philosophy of mind. Quite surprisingly, we were led to describe first the mental mechanisms behind *adherence*. Let's turn now to the issue of patient *nonadherence*.

Why does a patient not follow the advice of her doctor even when she realizes that not doing so is bad for her? She should do it, not only from the doctor's point of view (this is the classic definition of adherence) but also from *her own* point of view. Everyone involved, it seems, agrees that the treatment recommendation is a great idea—and yet the treatment never gets off the ground. Even though the patient has objective reasons to do it (she "sees something desirable in all actions that improve her health; she believes that losing weight is this type of action"),

© Springer International Publishing Switzerland 2015   89
G. Reach, *The Mental Mechanisms of Patient Adherence to Long-Term Therapies*,
Philosophy and Medicine 118, DOI 10.1007/978-3-319-12265-6_6

she still does not stay on the diet. The exasperated physician doesn't know what to make of the patient's behavior, which seems completely irrational. Some physicians distance themselves from these patients: "Come back when you're ready," they say.

What is so striking is that this exasperation is seldom lost on the patient, who is conscious of her behavior but does not understand it: When asked why she behaves in this strange manner, she does not try to hide the fact that she did not stick to her diet, adding with a sigh—"I know, I should, but I can't help myself."

A simple explanation is that the patient lacks knowledge of the diet and what it requires of her. She may not know how to identify which foods have high cholesterol content, or she may not know how to count her caloric intake. She might be confused because her physician told her to reduce her cholesterol intake, but on the TV news she hears that there is "good cholesterol" too.

We already recognized that the lack-of-knowledge explanation often falls short, as even patients who are well-educated about their medical conditions have problems of adhering to therapy. The doctor who smokes is perhaps the simplest demonstration of the insufficiency of the "knowledge is power" approach to treatment planning. A more complicated example is that of diabetic women who find out they are pregnant and suddenly begin adjusting their insulin doses (or quit smoking, or stop using cocaine). It often is not even necessary to remind them of the knowledge they already have.

As noted previously, other explanations have utility, at least sometimes: A hidden fear or other psychological issue, denial of illness, lack of money or other resources, and so forth. The previous section highlighted how poor habit formation and/or insufficient willpower and resolution can affect adherence to treatment over the long run. However, these explanations do not give a full description of the paradoxical nature of patient nonadherence: How is it *possible*? Again, understanding nonadherence calls for a philosophical interpretation.

Gary Watson has described several distinct situations that facilitate what seems to be an irrational behavior (Watson 1977; Smith 2003). Consider a person who takes one drink too many, and as a result is unable to drive her car responsibly. Watson distinguishes three explanations for how the drinker got into this situation. The first is *recklessness*: The drinker knows what she is doing, thinks that the value of the drink is sufficient to take the risk of not being able to fulfill her obligations, and accepts the consequences: Since she could have evaluated the risk differently, her behavior is reprehensible, even more so because it is not at all irrational. She has simply sized up the situation in a way which may lead to harm to herself or to others. The second explanation is *compulsion*: This time, the drinker knows she should not drink, but she cannot resist the compulsive desire of the forbidden drink. She can size up the situation any way she likes, but she's going to take the drink anyway.

Watson describes a third possibility: The person thinks it would be better not to drink, but in spite of this evaluation, *or, even as a result of this evaluation*, she drinks anyways. As opposed to the compulsive person, who is incapable of controlling herself, the third case is that of a person *who could have decided to act otherwise*, but did not. The reckless person throws caution to the wind, and drinks with a cavalier "to heck with it." The compulsive person drinks regardless of what she is thinking (or not thinking). Watson's third type drinks even though she knows it's a bad idea, and has the ability to refrain.

Consider the Chinese proverb describing the psychological progression of alcoholism: "At first, the man takes a drink; then, the drink takes a drink; then the drink takes the man." Perhaps these three stages are analogous to Watson's types? The "man takes a drink" is the reckless drinker; the "drink takes a drink" stage is when there is, as it were, a battle of wills between the man and the drink; and the final stage is the compulsive drinker, in which all willpower has been transferred from the man to the drink.[1]

Let us go back to patient nonadherence. We saw that not taking a pill, for instance, is a completely independent action. It is not the same as forgetting to take a pill; it is an action and not the absence of one. This has an important implication: If the patient knows, or rather thinks, that it would be better for her to take the pill (it is not a case of forgetting), and does not do it, then not only does she not perform the action she thinks is the best, but she also appears to choose the action she thinks is the worst, *when she could have performed the best action.* This description of patient nonadherence is strictly analogous to the third case in Watson's example of the person who takes a drink, where it is neither a decision of deliberate recklessness, nor a compulsion. As in the example of the person getting drunk, the choice is surprising.

6.1 Akrasia

The third type of person—the person who drinks despite her best judgment not to—Watson calls "weak." This perplexing human behavior has been described before: Aristotle called it *akrasia*. Other philosophers have used the term *incontinence*, or weakness of will. Aristotle tells us that Socrates rejected the possibility that a person could knowingly act bad:

> Socrates was entirely opposed to the view in question, holding that there is no such thing as incontinence; no one, he said, when he judges acts against what he judges best – people act so only by reason of ignorance.

He then adds:

> Now this view plainly contradicts the apparent facts, and we must inquire about what happens to such a man; if he acts by reason of ignorance, what is the manner of his ignorance? (Plato, Protagoras, 352b–358d; Aristotle, Nicomachaen Ethics, Book VII, 2, 1–2).

Aristotle devotes a large part of Book 7 of *Nicomachean Ethics* to *akrasia*, defined as a character trait which predisposes a person to intentional, non-compulsive behavior which at odds with his best judgment. The very existence of incontinent actions presents theoretical problems that have been analyzed by numerous

---

[1] I am also grateful to John Meyers for this comment.

contemporary philosophers, notably by Davidson, starting with the second Essay in *Actions and Events* and in *Paradoxes of Irrationality*; (Davidson 2001, 21–43; Davidson 2004, 169–189). There is also a recent collection of essays dedicated to it (Stroud and Tappolet 2003). The practical problems of incontinence have been grappled with by psychologists, theologians, physicians, law enforcement—indeed, any field which seeks to understand and shape human behavior.

6.2 Patient Nonadherence to Therapy as a Case of Akrasia

The concept of weakness of will—*akrasia*—may perhaps shed additional light on the problem of patient nonadherence.

Taking a medication is an action that can be considered desirable from the point of view of the physician who prescribed it, and to not take it is a typical example of nonadherence. But this concept of nonadherence is actually a reflection of *the physician's point of view*. On the other hand, the definition of an incontinent action is that the agent intentionally performs an action that *she herself* does not consider to be the best—not an action that is considered bad by another. In short, nonadherence is about disagreement between doctor and patient, whereas incontinence is about disagreement between the patient and herself.

If my physician has advised me to take some pills as part of my treatment and I don't do it, she can conclude that I am nonadherent to her advice. It may be that I do not think that taking the medication is, all things considered, best for me. I am refusing the prescription *against medical advice*. But if I do not take the medication even though I think it would be best to do so, my behavior is incontinent, akratic. Thus, to analyze the problem of nonadherence from the angle of *akrasia*, as I propose to do here, is to consider the patient's point of view. In other words, this analysis takes into account patient autonomy in the therapeutic relationship.

One may argue that perhaps "all things considered" is an unreasonable expectation: Who is ever in the position of being aware of all pertinent aspects of a situation, let alone having the time and ability to consider all of them? Even the doctor, whom we like to imagine knows all that is relevant in a medical situation, does not have complete information; and what information he does have may be weighted differently than how the patient (or another doctor) might weight it. For instance, a patient might not be worried about long-term complications in quite the same way that her doctor is; can we really say that this patient's choice to not adhere to treatment is akratic?

Nevertheless, we can assess a patient's choices in light of the information she does possess (regardless of whether it is less, more, or different than what her doctor possesses, whether it is biased by the use of heuristics—consider Tversky and Kahneman's theories 1974). When her choice is at odds with the information, we can rightly call it akratic.

Indeed, here is the tricky point: Deciding whether an action is akratic may be seen as a judgment, and many people, being sensitive about this sort of thing, may tell us: "Who are you to judge?" Thus, it is important to emphasize that akrasia is about internal incoherence, and not about whether the patient is doing or not doing what she "ought" to do according to others: The conflict is between the patient and her own judgment, between the patient and herself. Thus, akrasia steps us into the realm of meanings and belief, and loosens the exclusive coupling between "hard facts" and "rational decision making".

6.2.1 Philosophical Explanation of Akrasia

Incontinent actions are interesting from the philosophical point of view because they are difficult to understand. Davidson has analytically formulated the paradox of incontinent actions in three propositions (Davidson 2001, 23):

P_1: If an agent wants to do x more than he wants to do y and he believes himself free to do either x or y, then he will intentionally do x if he does either x or y intentionally.

P_2: If an agent judges that it would be better to do x than to do y, then he wants to do x more than he wants to do y.

P_3: There are incontinent actions.

P_1 and P_2 together entail that if an agent judges that the first action is better than the other, and if she believes herself to be free to choose, then she will perform the action she judges to be the best. But experience shows that this is not the case: P_3 affirms that there are truly incontinent actions. And that is why they are paradoxical: They should be impossible, since they contradict logic. Whence the typically philosophical question asked by Davidson in his 1970 paper: *How is weakness of the will possible?*

The first way to explain it is to hold that when the agent acts incontinently, she does not control herself, that she is possessed by unknown forces that keep her from acting in accordance with her best judgment. It is not she who acts, and thus it is not an action. Or that her judgment, under the influence of passion, pleasure or desire is so distorted that she is no longer capable of understanding that the action she is performing is bad. But, as Davidson notes, these explanations either deny that an incontinent action is possible (since in the first case it is not the agent who acts), or they deny that the agent has intentionally chosen the bad action having compared the relative virtues of the two. But as we have already seen, there are many instances of incontinent action (including the action of not doing something) where these explanations fall short.

6.2.2 A Choice Between Two Actions

We can make sense of incontinence by introducing two notions; conveniently, these notions complete the description of an action. Let us consider an example. Suppose I desire to lose weight, and I think that I can do so by exercising. And yet I do not exercise; I am 'incontinent'.

First, even if an action is considered in the form of a practical syllogism, it is obvious that this action would not exist in isolation. There would be in my mind many other syllogisms as well, and some would undermine the continent action. In our example, there may be something like this: "All action that assures my rest is desirable, I believe that not exercising is this type of action, so I do not exercise". This first explanation of the performance of an incontinent action can be seen as a moral conflict, a struggle between two opposing desires. But this explanation is not sufficient, because here, once again, the agent would not be responsible for her choice—she is merely the recipient of the outcome of the battle.

Davidson prefers another scenario, with three players: From the moment one has the choice between two actions, we can introduce a third component, which we can for now call the will; it decides in favor of one action or the other. It will choose the action with the strongest reasons (Davidson mentions "reason, morality, family, country" as reasons; for the case we are interested in, we can add the advice given by the physician) "based on all relevant considerations" (Davidson 2001, 35–36). But again, how can the agent rationally choose the one judged as the worst by reason? In other words, to use Davidson's title, how is weakness of the will possible? In the collection "Essays on Actions and Events", the essay on the weakness of will comes second, after the essay "Actions, reasons and causes", proving how crucial this question was for the author.

The second factor to be taken into account is the possibility that both actions may have their reasons 'for and against', i.e., neither action is *completely* desirable. Davidson suggests that there are two types of judgments. The first type of judgment is an unconditional judgment, always exact, a judgment *'sans phrase'*: It unconditionally leads to action. The second is a conditional judgment: It does not tell us whether an action is the best in itself, but rather whether it is the best taking into account the circumstances. These arguments cannot be properly assessed in a void; they can be understood only in a given context. Davidson calls this a *prima facie* judgment. For instance, a rule such as 'lying is wrong' might be judged to be inapplicable if the lie in question is uttered to prevent needless embarrassment. So this type of moral judgment does not unconditionally lead to action: 'So I do not lie'.

This can be analytically represented by an operator pf (x is better than y, r) where r is the reason why one thinks x is better than y. In this example, r is 'this particular lie, a white lie'. Here the conditional, *prima facie*, evaluation leads me to lying, when if I had stopped at the unconditional evaluation 'sans phrase', 'lying is wrong', it would have lead me not to lie.

As Ruwen Ogien notes,

> it is necessary [...] that I interrupt, in a second, my conditional (*prima facie*) evaluation, otherwise I would be completely incapable of acting. So in a way I fix my absolute duty, which is a rough evaluation, an unconditional judgment. The *prima facie* duty is what precedes this decision to believe that the action will be more just than unjust (or good rather than bad), a decision that transforms a *prima facie* or conditional duty into an absolute duty (Ogien 1993, 72).

Davidson thus proposes that reasoning that stops at *prima facie* or conditional judgment "is practical only in its subject, not in its issue" (Davidson 2001, 39) and might not result in a corresponding action.

To explain incontinent actions, Davidson comes back to the notion 'all things considered'. An action is said to be incontinent if it is performed while the agent has a better reason to do something else (Davidson 2001, 40).

> Definition: in performing x, the agent performs an incontinent action if and only if: (a) the agent performs x intentionally; (b) the agent believes there was another action she had the possibility of performing; (c) the agent judges that all things considered, it would be better to perform y rather than x.
>
>   This is the case for an action x that the agent performs for a reason r, but she has a reason r' that includes r, and on the basis of which she judges some alternative action y to be better than x. Inversely, we may say that an action x is continent if x is done for a reason r, and there is no reason r' (that includes r) on the basis of which the agent judges some action better than x. (Davidson 2001, 40)

This last point is important: The fact that the agent does not have a reason r' does not mean that no such reason exists. She simply knows of no such reason. This means that, first, mercifully, continent action is possible in the absence of exhaustive knowledge. And, second, it means that we are free to choose our motives, since *one* reason r is enough (there could be several reasons $r1$, $r2$). The patient does not need to know the physician's reasons; it is enough for her to have *her own* reason for exercising.

6.2.3   How Is Weakness of the Will Possible? The Principle of Continence

An incontinent act, then, simply fails to weigh the pros and cons of *all* the *available* arguments. Davidson compares this to the way we accept something as true:

> Carnap and Hempel have argued that there is a principle which is no part of the logic of inductive (or statistical) reasoning, but is a directive the rational man will accept. It is the requirement of total evidence for inductive reasoning: give your credence to the hypothesis supported by all available relevant evidence (Davidson 2001, 41).

As Ogien notes, this is equivalent to substituting an inductive image for a deductive model of action. More precisely, "the analogy of induction invites us to go back from the action to the intentions while the analogy of deductions suggests that we proceed from intentions towards the action." (Ogien 1993, 74–85). In the conclusion to his work, Ogien defends the idea that this change is present in Aristotle's work: *Akrasia* led Aristotle to abandon his deductive description of human action:

> Aristotle substituted an inductive theory of practical syllogism for the deductive theory that he usually defended. The action is no longer deduced from the practical syllogism: it is stated, directly identified. Then it is justified by more or less good reasons, the good being the universal ones, the bad, the particular ones. (Ogien 1993, 304–305)

Davidson suggests that similarly there is a principle of continence. When one abides by this principle, one commits to using all the available information before acting (this is the 'all things considered'), and to

> perform the action judged best on the basis of all available relevant reasons [...]. It exhorts us to actions we can perform if we want; it leaves the motives to us. What is hard

is to acquire the virtue of continence, to make the principle of continence our own. But
there is no reason in principle why it is any more difficult to become continent than to
become chaste or brave. (Davidson 2001, 41)

However, we must admit that the contents of our propositional attitudes, of what
we think, believe, hope, etc., have a certain consistency. Davidson proposes that
this consistency goes hand-in-hand with the principle of continence, and perhaps
the principle puts a premium on such consistency. This principle thus seems to be
a normative principle; in Davidson's words, it says:

> that one should not intentionally perform an action when one judges on the basis of what
> one deems to be all the available considerations that an alternative and accessible course
> of action would be better. This principle, which I call the Principle of Continence, enjoins
> a fundamental kind of consistency in thought, intention, evaluation and action. An agent
> who acts in accordance with this principle has the virtue of continence. [...] In any case, it
> is clear that there are many people who accept the norm but fail from time to time to act in
> accordance with it. In such cases, not only do agents fail to conform their actions to their
> own principles, but they also fail to reason as they think they should. For their intentional
> action shows they have set a higher value on the act they perform than their principles and
> their reasons say they should (Davidson 2004, 201).

And so, when the principle of continence fails, the agent is akratic.

6.2.4 An Incomplete Explanation

However, at the end of the first essay that Davidson devotes to this phenomenon,
he seems to conclude that it is impossible to understand a mind in the midst of
akrasia:

> If the question is read, what is the agent's reason for doing *a* when he believes it would
> be better, all things considered, to do another thing, then the answer must be: for this, the
> agent has no reason. Of course he has a reason for doing *a*; what he lacks is a reason for
> not letting his better reason for not doing *a* prevail. [...] But in the case of incontinence,
> the attempt to read reason into behavior is necessarily subject to a degree of frustration.
> What is special in incontinence is that the actor cannot understand himself: he recognizes,
> in his own intentional behavior, something essentially surd (Davidson 2001, 42).

6.2.5 Second Explanation: The Partitioning of the Mind

In *Paradoxes of irrationality*, Davidson goes further. He first recasts the problem
in these terms:

> What needs explaining is not why the agent acted as he did, but why he didn't act other-
> wise, given his judgment that all things considered it would be better. [...] A purely for-
> mal description of what is irrational in an akratic act is, then, that the agent goes against
> his own second-order principle that he ought to act on what he holds to be best, everything
> considered (Davidson 2004, 176–177).

For Davidson, the irrationality is in the agent's refusal to conform to this second order principle, the principle of continence, which states that one must act according to what one considers to be the best. In such a case, there occurs a striking phenomenon: A desire—the desire that causes the incontinent action—somehow impels the agent to ignore the principle. This interaction is surely causal, but it cannot be rational—how could it be, given that the effect is to *abandon* a principle of rationality?

We must then admit that

> in the case of irrationality, the causal relation remains, while the logical relation is missing or distorted. [...] there is a mental cause that is not a reason for what it causes. (Davidson 2004, 179)

To explain this phenomenon, Davidson evokes the partitioning of the mind:

> I went on to explain the vague and confusing notion of an attitude or principle being ignored or suppressed by again appealing to a Freudian idea, that of a partially partitioned mind. The idea, as I employed it, meant that attitudes in the same mind could be kept from actively interacting, so that the agent remained to some extent protected from the clash that would result from facing unwelcome thoughts or their consequences. (Davidson 1999, 404)

This hypothesis appears in his second essay devoted to *akrasia*, *Paradoxes of Irrationality*: First,

> there is a way one mental event can cause another mental event without being a reason for it [...]. This can happen when cause and effect occur in different minds (Davidson 2004, 180).

He gives as an example his desire to have the reader come into his garden: I grow a beautiful flower in order to attract you into it. You have a desire to see my flower and you enter my garden.

> My desire caused your craving and action, but my desire was not a reason for your craving, nor a reason on which you acted. (Perhaps you did not even know about my wish.) Mental phenomena may cause other mental phenomena without being reasons for them, then, and still keep their character as mental, provided cause and effect are adequately segregated.

Here, the dissociation is easy to explain: The cause and effect belong to two different minds. Davidson adds:

> But the way could be cleared for explanation if we were to suppose two semi-autonomous departments of the mind, one that finds a certain course of action to be, all things considered, best, and another that prompts another course of action (Davidson 2004, 180–181).

This division is

> necessary to account for mental causes that are not reasons for the mental states they cause. Only by partitioning the mind does it seem possible to explain how a thought or impulse can cause another to which it bears no rational relation (Davidson 2004, 184–185).

This essay is a superb defense of the fundamental concepts of psychoanalysis, and was published in a collection of essays on Freud's thought, and Marcia Cavell analyzed the relationships between Davidson's hypothesis of the portioning of the mind and the Freudian concept of irrationality (Cavell 1999, 407–423). However, although this concept is presented as a 'Freudian idea', Davidson stresses that this partitioning is not necessarily a partitioning between the conscious and the unconscious sectors, reminiscent of Freud's partitioning of the brain:

> the standard case of *akrasia* is one in which the agent knows what he is doing, and why, and knows that it is not for the best, and knows why. He acknowledges his own irrationality (Davidson 2004, 186).

It is as if the mind divides under the weight of two opposing arguments in one thought. This is analogous to the perception of "impossible objects". The Necker Cube is a simple and well known example, as is M.C. Escher's *Concave and Convex*. In these cases, an effort of the mind allows us to see an object in one configuration or another, but never both at the same time.

Davidson takes the supposition of a divided mind to be necessary for explaining an irrational mental cause. But this view can be criticized. Ruwen Ogien notes that,

> the description of the case proposed by Davidson prohibits us from attributing the accomplished action to the agent. The guilty party is the tempter-gardener [...]. The supposedly incontinent person is simply a victim. He has not even converted the temptation into actual beliefs and desires, because if this were the case, the reasons of the gardener would have become his own [...]. The supposedly incontinent person is manipulated from a distance, hypnotized, so to speak. He is an unaccountable sleepwalker, a passive toy at the mercy of foreign beliefs and desires [...]. He is no longer the author of his action; his behavior is in fact mechanical, sleepwalking, compulsive, not rational (Ogien 1993, 213–214).

Are these remarks pertinent if the process takes place in the same mind, as Davidson suggests? If this were the case, it would lead not to a performance of an incontinent action, but simply to compulsive behavior, and we would be back to the starting point in our analysis of incontinent actions. But it is precisely because the process takes place in the same mind that the argument of a "manipulation at a distance, of a hypnosis", falls through. It is in my mind that the two reasons to act coexist; the two domains of the mind, according to Davidson, are not separated but rather overlap; they are, theoretically, both accessible to my consciousness, and *I could examine their contents* if I activated my principle of continence. And while I "do not do the good that I desire to practice, but I do the evil that I do not want to do" as in Saint Paul's famous statement (Saint Paul, Romans), I may come to regret it once it is done, *and even before doing it, I may know that I will later regret it.* As Watson says,

> when one acts weakly, one wants to some degree to do what one judges best. Weakness of will is marked by conflict and regret (Watson 1977, 327).

Or more than regret, remorse: Baudelaire wrote (James McGowan Translation):

> While most, the rabid multitude of men
> Lashed by their Lust, in merciless torment
> Gather remorse on slavish Holiday,
> My Sorrow, take my hand and come away.

As Ogien notes

> to say that an action is incontinent is to interpret it in a certain way, by combining several
> perspectives. From the point of view of the agent, it comes down to adopting a certain
> reflective point of view of one's own actions. One never acts in an incontinent manner, but
> one becomes incontinent when, thinking about one's actions, the person judges that they
> could have been different and better. (Ogien 1993, 225)

However, the fact that there are incontinent action where one is conscious of
knowing, before doing it, that one is going to regret it, suggests that contrary to
Ogien's assertion, sometimes one truly acts incontinently. This strongly suggests
that incontinent action is a conscious phenomenon and is not simply a compulsive
behavior, but, in fact, an action.

6.2.6 Partitioning of the Mind and Patient Nonadherence

We must now ask whether the partitioned mind hypothesis is applicable to clinical
experience. Suppose a physician tells a patient that there is an obvious circum-
stance when she should adjust her insulin dose; for instance, her blood sugar was
very high every morning for the last two weeks, calling for an increase in dose.
When she is asked why she did not do it, two responses are common. Either she
cannot answer the question or she gives an incomplete explanation. The explanation is incomplete in that it applies only to one of the considerations pertinent
to the action. For example, she may say she is afraid of hypoglycemia, or, she is
afraid of gaining weight, or, lately she's had a lot of problems and didn't have time
to take care of her diabetes, etc. (Reach et al. 2005). The counterbalancing considerations, those that would motivate the adjustment of the insulin dose, seem to disappear from her practical reasoning, even though the patient is conscious of them.
This disappearance is, of course, irrational. There are pipe smokers who take the
warning labels out of the tobacco boxes so that they do not have to 'think' about it,
or perhaps because the labels confront their irrationality.

I hope you share my astonishment at this conscious aspect of incontinent
actions. Incontinent actions not only lead to later remorse, but they are *accompanied* by regret. An action that is perceived from the outside as an indulgence
is partly experienced as unpleasant from the inside. Knowing that the cigarette
at hand will plunge me back into the habit should spoil the very pleasure I seek
from it. Think of the miserable little handful of nuts that will ruin the entire
week's dieting efforts or the extra drink that will make me wake up tomorrow
bemoaning the weakness of my will. Loewenstein's "risk-as-feelings model"
(Loewenstein et al. 2003), pointing out the role of anticipated emotions in our
behaviors (see Chap. 3, p. 44), would suggest that anticipated regret should
protect me against incontinent actions. *Akrasia* becomes possible if anticipated
regret is inoperative.

Here I have assimilated the phenomenon of patient nonadherence to a particular case of *akrasia* (Reach 2000). We can find a rough draft of this description of the phenomenon at the very dawn of the Western thought in the writings of Democritus:

> Men ask for health in their prayers to the gods: they do not realize that the power to achieve it lies in themselves. Lacking self-control, they perform contrary actions and betray health to their desires (D/D B234)

and

> It is hard to fight desire; but to control it is the sign of a reasonable man (D/D B236).

And, as we saw earlier, Aristotle uses illness as an example of the consequences of incontinent actions. However, in the recent biomedical literature, it remains uncommon to draw on the literature on *akrasia* in order to explain the phenomena related to nonadherence. Authors seem to think that it is pertinent only to smoking and other problems connected to health behavior or addiction (O'Connell 1996; Heather 1998).

And yet patients themselves explicitly say: Excuse me, I can't help myself, I lack willpower. In a Spanish study, the reasons given for not eating a healthy diet were irregular hours at work (29.7 %), *lack of willpower* (29.7 %) and the suggested food not being very appealing (21.3 %) (Lopez-Azpiazu et al. 1999). Obese patients who ask for gastric banding say it very clearly: It will *make* me stay on the diet. Obviously, we could say, following Gary Watson (Watson 1977), that this is not weakness of will, but an irresistible desire, simply hunger, provoked by an increase of ghrelin, the hormone that is secreted by the stomach when it is empty and that stimulates hunger at the cerebral level.

Canadian women are similar in explaining why they don't exercise even though they know the benefits. 39.7 % say they don't have the time; 39.2 % say that don't have the willpower (Olmsted and McFarlane 2004). An English study of more than 10,000 participants has shown that those who mention internal barriers (I'm too busy, I lack the willpower, I'm too lazy) as opposed to external barriers (commute time is too long, I don't have enough money, etc.,) have significantly less physical activity (Ziebland et al. 1998). Could we say that they have an irresistible desire to rest? No, it is simpler to say that they lack willpower.

6.3   Another Medical Example of Irrationality: The Denial of Illness

Consider a phenomenon intuitively related to weakness of will: The denial of illness. The patient, although she knows deep down that she is ill, is convinced that she isn't. Denial is a normal phase of the psychological process that follows the announcement of a chronic illness, a process that is similar to mourning, mourning one's lost health. In this process, denial is short-lived, and followed predictably by other psychological responses. We know, from the description given by Gfeller and Assal, that the patient passes through several stages: After the *initial shock* comes a period of *denial*; followed by *revolt*, then the phase of *bargaining*, then a *meditative-depressive* state where sadness is predominant, and finally, *acceptance* of the illness and the constraints of its treatment (Gfeller et al. 1979; Assal et al. 1981). Normally, denial eventually gives way to the acceptance of illness, but with some patients the period of denial can persist. To take just one example, several years after the appearance of the illness, many diabetic patients tell us: I never could accept the idea that I am ill. Denial is defined by psychologists as

> a cognitive process that allows the coexistence of contradiction without the contradictions influencing each other. The contradictions are conscious, but the use of denial is not conscious. This process of disconnect between two contradictions is what gives denial its capacity for adaptation, allowing one to support reality in extreme situations (Spitz and Fischer 2002, 282).

The frequently cited belief that one is "immortal" is a generalized sort of denial. Alain Abelhauser notes that it is the nature of a human agent, on the intellectual level, to know on the one hand that we are mortal, but on the other hand, 'to ignore one's death', in other words, to imagine oneself, completely irrationally, as 'immortal', which of course determines in a certain way our integration into time. When we are healthy, we need this defense mechanism to protect us from the idea of death; such an idea may be unbearable if it were present. Experiencing others' deaths, or the occasion of a serious illness, typically brings mortality to conscious attention. Abelhauser adds in fact that

> serious illness, and the possibility of the fatal issue that it imposes, often challenges this type of mechanism by forcing the agent to confront the prospect of her own mortality and makes her reorganize her perception of time based on the amount she thinks is left. Such a reordering of temporal bearings is dangerous and painful, it condemns the individual to admit the uncertainty of her future and limits the possibilities for projection.

In redefining 'coping', "the ways available to the agent 'to deal with', to 'make do' with what is happening", he concludes that "it is not so much the question of *'dealing with'* as *'living with'*, and even *'durably living with'*." (Abelhauser et al. 2001, 82).

The denial of a serious illness could be seen as the persistence of the idea of immortality, like Caligula, who, mortally wounded at the end of Albert Camus' play, screams: I am still alive.[2]

Just as we could analyze patient nonadherence as a case of *akrasia*, we can relate the phenomenon of denial to another paradox of irrationality, self-deception: The agent does not believe what she should believe, despite all the available evidence (Mele 2001). However, there is a difference between the two phenomena. According to Davidson,

> self-deception and weakness of the will often reinforce one another, but they are not the same thing. This may be seen from the fact that the outcome of weakness of the will is an intention, or an intentional action, while the outcome of self-deception is a belief. The

---

[2] Or, perhaps just as denial of the inevitability of death is often a life-enhancing adaptation, so might be denial of death's cousin, illness. What is at issue clinically is when denial is used inappropriately. I am grateful to John Meyers for this remark.

former consists of or essentially involves a faultily reached evaluative attitude, the latter of
a faultily reached cognitive attitude. (Davidson 2004, 201)

Philosophers formerly analyzed self-deception as follows: The agent thinks that
(*p*) and *at the same time* thinks that (*non-p*). This is similar to the above definition
of denial given by psychologists: A "process that allows the coexistence of con-
tradictions". To explain this paradoxical situation, we can suppose that the agent
rids one of the beliefs from her consciousness; or that the agent has selected, particularly under the influence of desire, the argument that agrees with the proposition she wants to believe, in order to not believe the other (Mele 1992, 125–7).
This selection can also take place under the influence of an emotion, such as fear:
The agent believes that (*p*), not because she wants to believe that (*p*), but because
she fears that (*non-p*) may be true. As Mele notes, the interventions of desires and
fear is not necessarily self-exclusive, because the fear that (*non-p*) may in part be
explained by the desire that (*p*) (Mele 2001, 100).

Davidson has analyzed self-deception in exactly the same way as *akrasia*: Just
as there is a weakness of the will, there is a 'weakness of the warrant'. In the first
case, the agent does not conform to the principle of continence that tells her to
perform an action that, all things considered, she considers the best. In the second
case, the agent does not conform to the principle that asks her to consider all the
arguments before deciding whether a proposition should be retained. It is the principle established by Carnap and Hempel, and which Davidson took as his model
when proposing the principle of continence, whose deficiency explains *akrasia*.

One can see that Davidson's hypothesis concerning the partitioning of the mind
into autonomous regions can explain self-deception in the same way it did *akrasia*:

> The point is that people can and do sometimes keep closely related but opposed beliefs
> apart. To this extent we must accept the idea that there can be boundaries between parts
> of the mind; I postulate such a boundary somewhere between any (obviously) conflicting
> beliefs [...] It is now possible to suggest an answer to the question where in the sequence
> of steps that end in self-deception there is an irrational *step*. The irrationality of the result-
> ing state consists in the fact that it contains inconsistent beliefs; the irrational step is
> therefore the step that makes this possible, the drawing of the boundary that keeps the
> inconsistent beliefs apart. In the case where self-deception consists in self-induced weak-
> ness of the warrant, what must be walled off from the rest of the mind is the requirement
> of total evidence. What causes it to be thus temporarily exiled or isolated is, of course, the
> desire to avoid accepting what the requirement counsels. But this cannot be a *reason* for
> neglecting the requirement. Nothing can be viewed as a good reason for failing to reason
> according to one's best standards of rationality (Davidson 2004, 211–212).

Here we find an explanation that is strictly parallel to the one given by Davidson
for incontinent actions: The hypothesis of the partitioning of the mind allows, in
the same way, to understand that a desire can cause, without being its reason, the
rejection of a principle—the principle of the warrant in the case of denial, the principle of continence in the case of *akrasia*.

David Pears also uses the hypothesis of a multiple self to explain self-deception.
The agent can deceive herself because there are two agents inside her, the deceiver
and the deceived. If the agent can believe that (*p*) while she knows that (*non-p*),
it is because the deceiver in her, who knows that (not-p), manages to make the

deceived believe that (p). The deceiving sub-system of the mind might not be permanent; strikingly, Pears says it might be formed

like a camp set up for the duration of a particular campaign (Pears 1985, 77).

There is a difference between this type of partitioning of the mind and the Freudian division of conscious-unconscious. First, the latter represents a more structural, permanent organization of the mind. Second, the former is subject to the requirements of rationality, while the Freudian unconscious is not. But there are also similarities. Both fall under the influence of desires; and on both models of denial, the content of the active desire remains conscious even while its deceiving actions are not (Pears 1985, 77).

6.3.1   False Beliefs and Patient Nonadherence

To come back to nonadherence, we can easily imagine that a 'false' belief about one's illness can have a profound influence on a patient's behavior. It could lead her to form pro-attitudes toward apparently irrational actions: I believe that I am not ill; since I am not ill, I do not take the medication. But our analysis reveals that this may be merely apparent irrationality. If the belief is false, it is unsurprising that it should lead to an apparently irrational action. If Paul believes that he is not diabetic because he believes what a charlatan has told him, it is rational for him to refuse the insulin treatment. If Mary believes that insulin is a poison because she misunderstood what she saw on TV, it is rational for her to stop her treatment. But she may believe that insulin is a poison even though she was given ten arguments to show that this is false; in this case it is the selection of belief that is irrational, and this irrational selection is one of the mechanisms leading to self-deception. As we saw earlier, beliefs are born out of all the elements of the environment that an agent passively perceives, and it is this passive, involuntary character that makes them rational. It is when we *want* to believe, despite all the available evidence that we act irrationally.

6.4   Logical Mechanisms of Irrationality

Philosophy of mind distinguishes at least three types of irrationality: (1) Weakness of the will, which leads to the performance of actions described as 'incontinent' actions in which an agent acts contrary to her best judgment; (2) Self-deception; and (3) Wishful thinking. Let us now mark the differences between these last two.

It is important to note that, faced with a chronic illness, denial is a normal stage in the process leading to acceptance. This may suggest that denial of the illness, or, more generally, self-deception, results because desires and emotions *normally* intervene in the way we form our beliefs. Irrationality, then, is the 'reverse side' of the mechanism that allows for and simultaneously limits the enrichment of the content of the puzzle of the mind. Perhaps this is the case with each of these three types of irrationality, and in light of this we can now attempt their logical explanation.

In the case of *akrasia*, the agent performs an action that, all things considered, she should not have performed. It is the '*all*' of '*all*' things considered, that the agent has modified. This '*all*' is nothing other than the puzzle of her mind, and she has put aside the pieces that should have led her to perform the continent action. Davidson proposes that one can put aside the pieces by separating them into different partitions of the mind. In the model presented here, this means admitting that the mind is an aggregate of different puzzles.

But another explanation is possible: Christine Tappolet suggests that emotions influence our perception of values (she proposes that emotions are in fact the perception of values), and make akratic action understandable (Tappolet 2003, 97–120). The pleasure of smoking makes the fact that I smoke understandable. According to this model, *akrasia* would be the consequence of the mind's use of the normal role of emotions.

In the case of self-deception, the agent *wants* to believe something: For instance, that she is not ill, despite all the available evidence. Here, instead of a mind-to-world direction of fit she adjusts the world, so to speak, to suit her mind. Or more precisely, because she cannot truly change the world, and make it so that she is really not ill, she acts at the level of the puzzle of her mind. She accepts a skewed representation of the world by discarding unpleasant beliefs.

We saw that this logical error could be made under the influence of a desire (I want to believe that I am healthy) or of an emotion (I am afraid of being ill). In both cases, once again, the mind uses the normal roles of desire and emotion. And, as we have seen, these can have a positive effect: They can allow us to persevere in our actions and avoid procrastination; they can lead to a healthy revision of beliefs. But emotions can also lead to denial: Pierre Livet interprets self-deception as an agent's choice to make the revision that will cost her the least (Livet 2002, 96–97), and Vasco Correia proposed recently an emotional account of self-deception (Correia 2010).

Finally, in the case of wishful thinking, the agent comes to believe that her desire has come satisfied when in fact it isn't. She has not even tried to change the world by asking you to read her book (world-to-mind fit). No one has read her book, and yet, she believes that her book is being read. Here, it seems that the agent has mistakenly 'moved' the content of a desire from one attitude (a desire) to another (a belief). The desire modifies the structure of the mind, when desire's normal role is to modify the world. To mistake one's desires for reality is like living in one's dreams. In other words, the irrationality is to mistake one's representation of the world for the world itself.

Desires and beliefs are both reason and cause for an action. And this is true whether the relevant belief is true or false, whether the action is rational or irrational. This is important because it dissociates truth and rationality, and this in turn should put into perspective our judgments about another's rationality. An action that we judge to be irrational from the outside might be performed perfectly rationally from the point of view of the agent who has her own reasons. Patient nonadherence is diagnosed from the point of view of the physician, and it will always appear to the physician to be irrational if she judges it in light of her own references. If we are to truly understand nonadherence, we must consider the patient's multiple and complex points of view. It is when the agent performs a certain action rather than another, which all things considered, she should *herself* consider to be better that she manifests her irrationality.

Here, we reach the object of our research: *Mind and Care*. However, although the analysis presented here helps us to understand how *both adherence and nonadherence* are possible, we have now to explain why *some people* will be adherent to long-term therapies, and others won't.

I am a sick man. I am a wicked man. An unattractive man. I think my liver hurts. However, I don't know a fig about my sickness, and am not sure what it is that hurts me. I am not being treated, and never have been, though I respect medicine and doctors. What's more, I am also superstitious in the extreme; well, at least enough to respect medicine... No, sir, I refuse to be treated out of wickedness. Now, you will certainly not be so good as to understand this.

These are the first lines of Dostoyevsky's Notes from the Underground (Dostoyevsky 1994). Can we go further in trying to understand this irrationality, or what Dostoyevsky calls *wickedness*?

References

Abelhauser A, Lévy A, Laska F, Weill-Philippe S. Le temps de l'adhésion. In: L'observance aux traitements contre le VIH/SIDA. Agence Nationale de Recherche sur le SIDA; 2001.

Aristotle. Nicomachaen ethics. Book VII, 2, p. 1–2.

Assal JP, Gfeller R, Kreinhofer M. Les stades d'acceptation du diabète. Journ Annu Diabetol Hôtel Dieu. 1981;223–35.

Baudelaire C. Meditation. The flowers of evil (trans: McGowan J). Oxford World's Classics; 2008.

Cavell M. Reason and the gardener. In: Hahn LE, editor. The philosophy of Donald Davidson. The library of living philosophers. vol. XXVII. Open Court; 1999, p. 407–21, and Davidson's response, p. 422–23.

Correia V. La Duperie de soi et le problème de l'irrationalité: Des Illusions de l'esprit à la faiblesse de la volonté. Éditions Universitaires Européennes; 2010.

Davidson D. Actions, reasons and causes. J Philos. 1963;60:685–700 (Reprinted in: Essays on actions and events. Oxford: Clarendon Press; 2001).

Davidson D. Two paradoxes of irrationality. In: Wollheim R, Hopkins J, editors. Philosophical essays on Freud. Cambridge: Cambridge University Press; 1982, pp. 289–305 (Reprinted in: Problems of rationality. Oxford: Clarendon Press; 2004).

Davidson D. In: Hahn LE, editor. The philosophy of Donald Davidson. The library of living philosophers. vol. XXVII. Open Court; 1999.

Dostoyevsky F. Notes from the underground (trans: Pevear R, Volokhonsky L). Vintage; 1994.

Gfeller R, Assal JP. Le vécu du malade diabétique. Folia Psychopractica, 10, Hoffmann-Laroche et Cie, SA Bâle; 1979.

Heather N. A conceptual framework for explaining drug addiction. J Psychopharmacol. 1998;12:3–7.

Livet P. Émotions et rationalité morale, P. U. F., Collection Sociologies; 2002.

Loewenstein G, Read D, Baumeister RF, editors. Time and decision, economic and psychological perspectives on intertemporal choice. New York: Russell Sage Foundation; 2003.

Lopez-Azpiazu I, Martinez-Gonzalez MA, Kearney J, Gibney M, Martinez JA. Perceived barriers of, and benefits to, healthy eating reported by a Spanish national sample. Public Health Nutr. 1999;2:209–15.

Mele AR. Irrationality, an essay on akrasia, self-deception and self-control. Oxford: Oxford University Press; 1992.

Mele AR. Self-deception unmasked. Princeton: Princeton University Press; 2001.

O'Connell KA. Akrasia, health behavior, relapse and reverse theory. Nurs Outlook. 1996;44:94–8.

Ogien R. La faiblesse de la volonté, P.U.F.; 1993.

Olmsted MP, McFarlane T. Body weight and body image. BMC Women Health. 2004;4(Suppl 1):S5.

Pears D. The goals and strategies of self-deception. In: Elster J, editor. The multiple self. Cambridge: Cambridge University Press; 1985.

Plato. Protagoras, 352b–358d.

Reach G. Application de la théorie causale de l'action à l'analyse de la non-observance thérapeutique. Presse Med. 2000;29:1939–46.

Reach G, Zerrouki A, Leclerc D, d'Ivernois JF. Adjustment of insulin doses: from knowledge to decision. Patient Educ Counsel. 2005;56:98–103.

Saint Paul. Romans. 7:15–25.

Smith M. Rational capacities, or: how to distinguish recklessness, weakness and compulsion. In: Stroud S, Tappolet C, editors. Weakness of will and practical irrationality. Oxford: Clarendon Press; 2003.

Spitz E. Les stratégies d'adaptation face à la maladie chronique. In: Fischer GN, editor. Traité de psychologie de la santé. Dunod; 2002.

Stroud S, Tappolet C, editors. Weakness of will and practical irrationality. Oxford: Clarendon Press; 2003.

Tappolet C. Emotions and the intelligibility of akratic actions. In: Stroud S, Tappolet C, editors. Weakness of will and practical irrationality. Oxford: Clarendon Press; 2003.

Tversky A, Kahneman D. Judgment under uncertainty: heuristics and biases. Science. 1974;185:1124–31.

Watson G. Skepticism about weakness of will. Philos Rev. 1977;86:316–39.

Ziebland S, Thorogood M, Yudkin P, Jones L, Coulter A. Lack of willpower or lack of wherewithal? 'Internal' and 'external' barriers to changing diet and exercise in a three year follow-up of participants in a health check. Soc Sci Med. 1998;46:461–5.

Chapter 7
Time and Adherence: A Principle of Foresight

Abstract When we consider two possible and contradictory actions, their adjacent desires have corresponding rewards, which might not be available at the same time. Thus time influences our decision, leading to what is described as intertemporal choice: Often, in chronic diseases, the choice of adherence or nonadherence can be seen as a choice between an abstract and distant reward, maintaining health, and a near and concrete reward, for example the pleasure of smoking. Many people are naturally impatient, preferring a small, near reward to a large, distant reward. This trait, patient or impatient, may be linked to adherence. I propose that, in the particular case of akrasia represented by patient nonadherence to long-term therapies, there is such disequilibrium between the two types of actions, the continent and the incontinent, that it does not allow the principle of continence to express itself, or rather this principle becomes insufficient, or even inappropriate, if used alone. This leads me to propose a hypothesis introducing a second principle, which I call the principle of foresight, which pushes us to give priority to the future, i.e., to accept taking care of ourselves. Maybe we have here something that is essentially human. One can also speculate that this differentiation is accomplished slowly in adults, leading from the simple age of reason to an age of foresight. According to this hypothesis, not conforming to this principle leads to nonadherence.

7.1 The Effect of Time

In the first part of this book, we saw that the phenomenon of patient nonadherence occurs especially in the case of chronic disease. Chronic implies, of course, an illness of long duration, and in particular, of endless duration: The patient will have to deal with treatment indefinitely. When a diagnosis of a chronic illness is experienced by the patient, it is easy to imagine that such news will cause an upheaval in the patient's life: Nothing will be quite the same ever again.

© Springer International Publishing Switzerland 2015

G. Reach, *The Mental Mechanisms of Patient Adherence to Long-Term Therapies*,
Philosophy and Medicine 118, DOI 10.1007/978-3-319-12265-6_7

Time—especially as it is experienced psychologically—has been omnipresent in this study. It intervenes first in the *process* that leads from the initial denial to the acceptance of the illness; this process takes time, and cannot be rushed through. Time appears again in the numerous behavioral models we reviewed in the first section of the book. In the *Health Belief Model, at a certain moment,* the occurrence of a signal might be necessary for the patient to decide to begin treatment. But it is especially Prochaska's *Transtheoretical Model of Change* that explicitly describes the role of time in the adoption of a health behavior, and we saw how these different stages could be interpreted in terms of intention, decision and resolution. These mental states situate "therapeutic agency" in the context of a project, i.e., taking time into account.

7.1.1 Time and the Choice Between Two Desires

When we consider two possible and contradictory actions, their adjacent desires have corresponding rewards, which might not be available at the same time. We shall see that our perception of the respective importance of the rewards may vary over time. Thus time influences our decision, leading to what is described as *intertemporal choice*.

It's obvious that from the economic point of view, the worth of goods declines over time: In five years, my car is not going to be worth as much as it is today. In economics, this loss of value over time is often exponential: Value diminishes exponentially when it loses a set percentage of value in each unit of time. For example, a new car loses, say, 20 % of its value with each passing year, so that a new car costing $25,000 will be worth $20,000 at the end of the first year, then $16,000 and the end of two years, and so on.

But in addition to this depreciation simply due to time, there is another type of depreciation, a psychological depreciation. I can prospectively *evaluate* it, so it is also connected to economic considerations: If I am promised a *new* car in five years, I will not assign to it the same worth as if I'm promised the car in a month. I take into account the fact that it is not certain that the car will still be available in five years, that its accessories will not be the most up to date, that the promise might be broken, or even that I might not live that long, etc. If I conclude that it's the same to receive $100 now as to receive $120 in a year, this means that my time discount rate for money is 20 % (Chapman 2003, 395). In counting the future value of goods, I take into account not only my estimate, today, of the value that it will have at that time, but also the probability that I will in fact be able to enjoy it. The further removed the award, the smaller the probability that I will actually get to enjoy it. This 'psychological' discounting does not depend only on objective criteria: The people who have a strong tendency to discount the value of goods over time are those who, being impatient, are also those who are less capable of postponing their enjoyment. The impatient people might also ask themselves: What's the relationship between the person that I will be in ten years and the one I am today; should I deny myself pleasure today for the sake of this person, who maybe won't have that much in common with me?

However, Shane Frederick, analyzing this idea expressed by Derek Parfit, did not find, in an experimental study, a correlation between an estimation of the degree of connection between the now-me and the future-me and the discount rate of the future (Frederick 2003). At the end of our analysis we will see that Parfit himself recognizes that this position might not be moral.

So, psychologically, the value of goods declines over time, though the decline is not typically exponential, because the value of the discount rate is not constant.

Indeed, the discount rate is higher when considering the near future as opposed to the distant future. For example, people were asked how much money they would agree to receive one month hence, one year, or 10 years in exchange for $15 now. The answers were respectively, on average, $20, $50 and $100. If we use the exponential model, these results give discount rates of 345, 120 and 19 % for the three periods (R. Thaler, quoted in Frederick 2003, 25). George Ainslie was the first to demonstrate that the mathematical function that best describes this type of curve is the hyperbolic (and not exponential) function (Fig. 7.1) (Ainslie 1985, 140, 1992). He suggested that this follows from a rule, proposed by Richard Herrnstein in 1961, according to which animals who have to choose between two behaviors spend an amount of time that's inversely proportionate to the delay before the occurrence of the consequences.

Many types of functions have been described, for example $D(t) = 1/t$, $D(t) = 1/(1 + kt)$, etc., (Frederick 2003, 69, Rachlin et al. 1991) where $D(t)$ represents the evaluation of the value of goods over time t. The hyperbolic nature of the function of discounting makes the discount rate for the nearest future the highest. Experimentally, one can show that for humans, it is during the first year that the discount rate declines the most: When one considers rewards more than a year

Fig. 7.1 The difference between hyperbolic (*solid line*) and exponential (*dashed line*) discounting

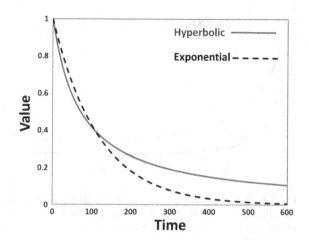

from the present, the rate is nearly constant. This is consistent with the way Jon Elster explains the hyperbolic version of the discounting function:

> Perhaps the central intuition behind this view is that individuals have a strong preference for the present compared to all future dates, but are much less concerned with the relative importance of future dates. If they receive a big sum of money today, for instance, they may decide to spend half of it immediately and allocate the rest evenly over their lifetime (Elster 2000b, 25).

This idea is remarkably close to the way Spinoza saw the effect of time on the strength of passions:

> From our note to Definition 6, IV, it follows that with regard to objects that are distant from the present by a longer interval of time than comes within the scope of our imagination, although we know that they are far distant in time from one another, we are affected towards them with the same degree of faintness. (Spinoza, *Ethics*, book IV, scholium of proposition 10).

In other words, the curve describing our valuation of goods over time, between the present moment and the moment of its acquisition, becomes very concave as it gets to the top. So the estimated value rises but slowly as we get closer, but when we are very near to the goal the value rises rapidly.

We can also illustrate this hyperbolic nature of the function describing the strength of desire over time by saying that this strength abruptly rises, almost asymptomatically, when one gets close to the reward. Think of the night before the delivery of a long-awaited car that the impatient individual ordered several months ago (Fig. 7.2).

It is possible to determine the value of the discounting parameter k, used in the equation: Experimental studies show that the parameter is 0.77 and 0.16 s^{-1} for pigeons and rats, respectively.

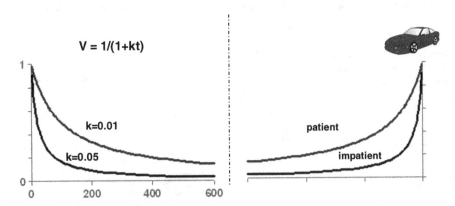

Fig. 7.2 Hyperbolic *curves* for a patient (k = 0.01), and an impatient (k = 0.05) individual. The night before the delivery of the car, the force of the desire will increase more abruptly in the second one. Actually, *curves on the right part of the figure* were constructed from the *left curves* by a rotation around a vertical axis

They are thus capable of postponing the acquisition of goods a tiny bit (this type of experiment involves offering the animal a small or a large quantity of food, the latter only being given after a certain delay, which is then varied to determine at which point the animal will choose the smaller quantity), while humans (the experiment generally consists of offering different sums of money, available after different delays) are capable of delaying gratification 0.014 months^{-1}. These results suggest that humans have developed the capacity for patience to a degree vastly superior to that of birds and animals (Bickel and Johnson 2003, 423). It can also be shown that smokers discount money over time at a higher rate than non-smokers (Bickel and Johnson 2003, 429) and that the discount rate is higher for health than for money. This last observation could partly explain the difficulty of adopting and persevering in a health behavior (Chapman and Elstein 1995); although, as we shall see, this conclusion needs to be revisited.

7.1.2 Intertemporal Choice Between Two Rewards

What matters most when choosing between two rewards, one imminent and the other distant? The critical factor is one's evaluation, *at the moment of choice*, of their respective values: This is what is going to give the motivational force for obtaining one or the other. As the saying goes, *a bird in the hand is worth two in the bush*. It is because we have performed this calculation of probability that we may prefer a small reward that is near (*all of a sudden it seems* very important) rather than a larger, distant award (it *seems* to be of a lesser value). Let us refer to this evaluative factor as *time preference*.

7.1.3 The Concept of Preference Reversal

If one finds herself removed from two rewards, the more distant reward might seem more important than the closer one, because the latter is still far away. But because of the hyperbolic, rather than exponential, nature of discounting, there is a moment when, as the reward nears, its motivational force abruptly rises and may become superior to that of the distant reward. And so the agent chooses it, even if beforehand its value appeared inferior to the one that the distant reward would have in the end. Figure 7.3 gives a graphic representation of this phenomenon.

Let us imagine an individual who is advancing toward two rewards: One is near while the other is more important, but further away, and she must choose: It is one or the other. She will schematically pass through three stages. In stage A, she evaluates the more distant reward as being more important than the more immediate reward: She prefers the distant reward. But as she approaches the more immediate award, she enters stage B: Because of the hyperbolic nature of the discounting curves (this would not be the case if the discounting function were exponential),

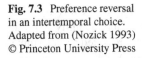

Fig. 7.3 Preference reversal
in an intertemporal choice.
Adapted from (Nozick 1993)
© Princeton University Press

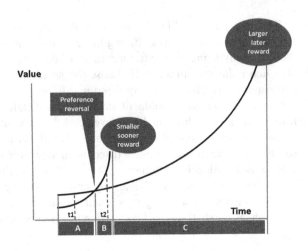

the discounting curves cross and as the value of the nearer reward acutely rises, and the nearer reward now appears to be more important than the distant reward. There is a real reversal of preferences; the individual now prefers the immediate reward: And so she chooses it rationally, even though it has an objectively inferior value compared to the distant reward—and in spite of her having once preferred the more distant reward.

The following metaphor, taken from Elster, illustrates the phenomenon of preference reversal: Imagine standing in front of a small building that has a taller building behind it. As you approach the first building, it will at some point obscure the taller building behind it, as if it were bigger. So it is with our preferences: Rewards which are nearer loom larger in our preference calculation.

Elster also helps us to understand preference reversal due simply to the passage of time (Elster 1999b, 430). When I enter a restaurant (period A) my desire to stay on a diet may be stronger than my desire to eat a dessert. Unfortunately, as the dinner progresses, my good intentions weaken, perhaps also due to the consumption of alcohol, and when the waiter arrives with the dessert cart, my desire for cake suddenly grows and becomes superior to my desire to stay thin (period B). We must stress, as Elster does, that here it is not the sight of the dessert that propels the force of desire, but simply the passage of time: One can say that the server and the cart arrive at the right time. We might note here in addition that it is not by chance that as you arrive at the restaurant, you are offered a cocktail before you can even look at the menu. Alcohol increases the discounting of values over time and the resulting 'myopia' makes you see everything on the same plane (Steele and Josephs 1990).

For Damasio, in *Descartes'* Error,

> this concept that has been proposed to explain the behavior of individuals under the influence of alcohol and other drugs. Inebriation does narrow the panorama of our future, so much so that almost nothing but the present is processed with clarity (Damasio 1994, 218).

Then, when you take the menu, the curves of your conflicting desires—one to eat cake and one to stay thin—will have already crossed and you will order the dessert without hesitating, postponing your diet and your good intentions until tomorrow.

Robert Nozick pertinently asks why we have a tendency to call stage B the temptation stage, and why we think it must be resisted (Nozick 1993, 16). Why do we consider that it is during stages A and C, when the more distant reward *appears* to be the more important one, that the agent should make her choice, even though, during stage B, *it appears to be rational for the agent, taking into account her preferences at that moment*, to choose the more immediate reward? Nozick offers the following explanation: Stage B is not an appropriate period for a decision because it is not representative of what the agent generally believes: It is very short, in fact a lot shorter than the sum of stages A and C. Let us suppose that the agent has consumed the immediate reward that keeps her from receiving in the future the more distant reward. Afterwards (at the end of stage B), she will evaluate the reward she has now given up, and she will again assign it a greater value because she is now at stage C (she again can see the taller building!), and she will regret her decision. But if she has resisted temptation at stage B, then when she arrives at stage C and looks back, she will still believe herself to have made the right decision.

An ex-smoker wrote me that he keeps a pack of his favorite cigarettes in his desk drawer and from time to time looks at it "like the toreador at the bull". This extraordinary flirtation with risk is only possible, in the long run, if the former smoker is already in period C, where the contemplation of the unsmoked cigarettes confirms the validity of his choice.

So the reason we believe the agent should resist temptation during stage B is that it is relatively short, and the choices made in stage B are marked by regret, while those in A and C are not. Besides, during period B, the agent is at periods A or C for other similar choices, which may also influence her.

And sometimes one does resist temptation. Indeed, we saw that doing so requires a particularly human capability. The rewards expected at different times can be chosen for their importance, but on condition of making the choice sufficiently in advance, i.e., at the moment t1 in Fig. 7.3 where, because of the hyperbolic nature of discounting curves, the attraction of the distant choice appears to be even more important than that of the more immediate reward. It is enough for the agent to realize that there is a risk of the two curves crossing: After this moment (when it reaches t2), the attraction of the more immediate reward will be stronger and it will be too late. Before the curves cross, the agent will use her self-control, in the form of a ruse, to avoid giving into temptation: A system of warnings, or what Ainslie (1985, 144–145) and Elster (2000b) call a precommitment device or strategy. For instance, Ulysses faces two choices: The distant reward, returning to Penelope; the immediate reward, a dalliance with the Sirens. Before the desire for the Sirens becomes too strong, he appeals to his self-control and uses the well known ruse: He commands his sailors to tie him to the mast. And knowing that his will is not all-powerful, he *then* instructs them:

> If I beg you to release me, you must tighten and add to my bonds. (Homer 2003, *Odyssey*, Book 12, quoted in Elster 1977)

Implemented at stage A, Ulysses' ruse keeps him from giving into temptation during stage B.

If medical nonadherence depends on similar mechanisms, these considerations should be useful for developing ways to improve adherence.

7.1.4 The First Solution: Precommitment Strategies

According to Elster, one cannot voluntarily change one's rate of discounting for future rewards if it's elevated. Rather, *wanting* to be motivated by long-term consequences entails that one already is motivated by such rewards. Compare morality: If one wants to be immoral, then one *is* immoral (Elster 2000b, 28). Let us call the failure to be motivated by far-removed consequences *impatience*; then we may say that one either is or is not impatient. Elster and Skog note that desires, which are the driving force behind our actions,

> cannot be classified as rational or irrational [...] This argument applies not only to substantive preferences for one good over another [...] Some like chocolate ice cream, whereas others have a taste for vanilla. [...] Similarly, it is just a brute fact that some like the present, whereas others have a taste for the future. If a person discounts the future very heavily, consuming an addictive substance may, for him, be a form of rational behavior. (Elster 1999a, 17)

Inasmuch as this type of impatience leads to the possibility of 'caving in' associated with a reversal of preferences similar to the one described above, the first way to avoid it is to develop the techniques of precommitment described by Elster (2000a, 188–190). For example, one might not keep cigarettes—or sweets—in the house to render impulsive use impossible. Such a strategy may also force some delay between the urge and gratification; while driving to the store to buy some cigarettes, one has a moment to re-think the decision. Another precommitment strategy is to associate a punitive cost with the reward. For example, the drinker who takes Antabuse has, in effect, set it up so that alcohol consumption is inextricably linked with severely unpleasant physical sensations. Yet another strategy is to increase the value of the more distant reward: "If I can lose some weight, I will be able to wear my favorite suit again, something that's much more attractive to me than the abstract prevention of cardiovascular diseases". This increase in value can occur by cognitive reframing (as in the previous example), or by even attaching an additional reward to the original goal: "If I lose 30 lbs and can fit into my favorite swimsuit, I'll go on a vacation to the beach."

Nozick suggests that using a principle is a way to fight temptation (Nozick 1993, 17–20). In order to not start smoking again, I tell myself: *I will never again smoke a pipe*. The essence of a principle, for Nozick, is that it confers on the act of smoking *this particular pipe* the symbolic value of *all future pipe smoking* (Nozick 1993, 26). Speaking of this value of principles, we can mention here what are called New Year's resolutions. Making them on this particular day makes them more effective, thanks to its symbolic strength. A study showed that the people who follow this tradition are more adherent to their commitments to lose weight, exercise or quit smoking than people who have the same objective at a different time of the year (Norcross et al. 2002), and a Canadian article linked this to the increase in cigarette advertisements in January and February, the goal being to counter these good resolutions, which are unfortunate for cigarette manufacturers (Basil et al. 2000).

This explains why using a principle allows one to resist temptation during stage B: *This pipe* that I run the risk of smoking becomes, symbolically, charged with all the future misdeeds of tobacco use, and thus its value cannot go up enough to make me cave in. And this is also a way to avoid a relapse when I have arrived at stage C for the intertemporal choice concerning *that pipe*. Violating my principle would symbolize the destruction of all the efforts I have made in refusing all of the preceding pipes. Hence, the sentiment "I've made it this far…" can serve to motivate us during difficult stage B temptations.

The use of a principle as described by Nozick is analogous to what George Ainslie calls a personal rule (but it is only analogous: We will see at the end of this chapter the fundamental difference between the two concepts). An individual bundles her temptations into a group, so that each choice involving a temptation becomes a precedent allowing her to predict all the future choices in this group. Thus, she becomes capable of emphasizing her desire to attain an important but distant goal each time she finds herself faced with a small step that would take her in the wrong direction (Ainslie 1985, 145). Ainslie suggests that this tactic is the same, at the moment of choice, as considering the sum of future choices. The respective curves of the two rewards, the shorter-smaller one (for example, the dessert), and the larger-later one (staying thin, for instance) are added together and end by taking on a quasi-exponential form that prevents them from crossing (Fig. 7.4).

On the left side of the figure (a), we can see someone headed towards the choice between a cake and the objective of staying in good health. The hyperbolical form of the curves describing the force of the respective desires makes

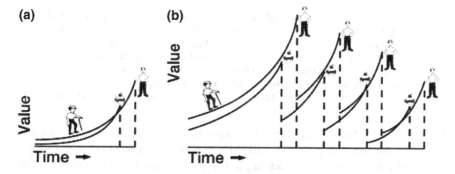

Fig. 7.4 The personal rule according to Ainslie (2006). Originally printed in (Ainslie 2006), © MIT Press

them cross, and it becomes rational for the agent to order the dessert. On the right side of the figure (b), the agent believes that the choice she makes today commits her to future choices, and she adds the respective curves: They take on a form that does not allow them to cross.

7.1.5 Second Solution: Intermediate Rewards

In health maintenance, for example in the case of cardiovascular diseases, a major obstacle is that the efforts needed to maintain health are prolonged and (often) without visible benefit.

As a doctor, I believe we have here a key to improving patient adherence. Perhaps it is more effective to tell the person to whom you are prescribing a diet that, thanks to her efforts, she will simply *feel better* quickly or that she will be proud of committing herself to a health objective. To someone who wants to quit smoking, knowing that she will be able to go up the stairs without getting out of breath, or that she will no longer cough when she wakes up each morning, or seeing the relief on the faces of her spouse and children when they see her improved health are all more immediate and meaningful rewards. The doctor need not brandish the ghost of diseases the patient maybe has never heard of and that often remain just abstract scarecrows (perhaps measuring glycated hemoglobin—which reflects glycemic control in the diabetic patient over the preceding three months—also has an additional value: It acts as an intermediate reward).

This approach calls for strategic use of emotions. Pierre Livet suggests that emotions may be another strategy for guiding and motivating behavior (Livet 2002, 80), as compared to precommitment strategies, which are not foolproof. Because rewards are in large measure emotional, it matters very much whether we experience an emotion now versus experience it in the future. In the example of the dinner, it is immediate pleasure of a piece of cake as weighed against the unpleasant emotions that I will feel tomorrow when I step on the scale and find out that I have ruined the dieting efforts of an entire week. Livet:

> my emotions can lead me to the right path, because they oppose, maybe not the reality, but at least the impact of reality on my affects to the desires that don't take them into account. Emotions are a reminding force for the desires [...]. Nonetheless, emotions can have a reminding force, while the desires cannot. And Elster's objection that all that passion can do, reason will do just as well with less effort, does not apply here, because long-term reason does not have the force of reminder for short term reason, at the moment when the decision is made. The differential of emotions can, on the other hand, play this role.

We can suggest that a therapeutic technique to be developed could specifically consist of helping the patient identify the intermediate rewards that she imagines being able to obtain by accepting a treatment regimen, and to also explicitly address the emotions related to the alternatives she faces.

Olav Gjelsvik suggests that

> a simple form of self-control of one's mind in connection with quitting might be to concentrate on all the good things that accompany nonsmoking [...] The result of the concentration might be a change in the present assessment of the value of nonsmoking compared with smoking. (Gjelsvik 1999, 57)

Obviously these 'good things' must be emotions. We saw that the concept of resolution implies that the agent is able to mentally repeat the value of the decision in the long run. This must be done in the most attractive way and Baumeister suggests that the vividness with which paradise and hell were represented in medieval imagery was not accidental (Baumeister and Voes 2003, 210].

In Alcoholics Anonymous, drinkers are advised to 'think through the drink', meaning to imagine the emotions they will experience if they pick up a drink, how they will feel about the reactions of others, etc.[1]

7.1.6 A Criticism of the Notion of Incontinent Action?

Elster notes that it is debatable whether we should call the "caving in" during period B an incontinent action, because I chose to perform the action that I considered *at that moment to be the best* (point t2 in Fig. 7.3). It is only incontinent in relation to my general resolution of staying on a diet, which was still there when I entered the restaurant. In this case, it is strictly speaking an incontinent action only if the moment when the individual decides to perform the more immediate action, and the one that doesn't seem the best to him, is situated at t1 in Fig. 7.3, i.e., before the reversal of preferences. And in fact, taking up the example of the restaurant, Elster also notes that this crossing is not necessarily related to the effects of alcohol or to seeing the dessert cart. It may have already occurred in the taxi on the way to the restaurant (Elster 1999a, 22). This last remark could lead one to question the very existence of incontinent actions (Elster 1999b, 425–445).

Here it is necessary to note the difference between two types of phenomena that are called patient nonadherence: On the one hand are the actions involving non-renunciation of pleasure, for instance staying on a diet (it is significant that Elster uses the example of giving up on a diet at the end of dinner) or refusing a cigarette. On the other hand are the various tasks that the patient must perform in the course of the treatment of a chronic disease: We have argued from the start of the book that to refuse a cigarette is not the same thing as to take a pill.

First, while there is a strong correlation between addictive behavior and an elevated discount rate, just as the Theory of Interpersonal Choice predicts (Bickel and Johnson 2003, 419–440; Perry and Carroll 2008, for a full discussion see Story et al. 2014), the correlation may be weaker in other health behaviors: We observed in a small number of type 2 diabetic patients an association between preference for

---

[1] I am grateful to John Meyers for this remark.

a smaller-sooner monetary reward and adherence to medication and metabolic control (Reach et al. 2011) and in the study in obese diabetic patients, already quoted, where we demonstrated the association between seatbelt behavior and adherence to medication, we also observed that giving the priority to the future was an independent determinant of adherence to medication (Reach 2011). However, another study found only a weak relationship between the discount rate and such diverse health activities as exercise, weight loss, dental care, and seatbelt use (Chapman 2003).

Second, as Peter Herman and Janet Polivy show, it might be going a little too fast to take the so often used example of caving into eating dessert as the quintessential preference change described by the Theory of Intertemporal Choice. There are important differences between having to choose between $100 now and $200 next year and a dessert now and the possibility of staying thin: For one thing, it is not certain that I will in fact become thin if I do not eat this dessert (and previous unsuccessful attempts may reinforce this impression). In addition, for the monetary choice, you are offered a certain sum today and a larger sum tomorrow, while in the case of a diet, the future can be summed up as no cake today and still no cake tomorrow.

Nevertheless, the perception of time presumably plays a role in the decision to stay on a diet: Herman and Polivy consider that individuals who have a short life expectancy in front of them should be less inclined to stay on a diet (although they do mention the case of a man sentenced to death who for his last meal ordered a glass of Diet Coke! It is difficult to understand this except as the result of habit or of supreme irony). They admit therefore that experimental research is needed to clarify the relationship between perception of time and adhering to a long-term goal (Herman and Polivy 2003, 459–489).

Finally, Olav Gjelsvik discusses in detail these two conceptions of the weakness of will, Davidson's version that includes the phenomenon of *akrasia*, and that of Ainslie and Elster, which brings into play the reversal of preference (Gjelsvik 1999, 47–64). He shows that the two conceptions are not mutually exclusive; but he admits that there are at least certain cases where Davidson's explanation is more apt, such as when an individual falls back into a *bad habit*. Consider, for example, consciously accepting a cigarette, even while you are well aware *as you are lighting it* that *this cigarette* might be the event that will make you relapse, with serious consequences for the future.

Inasmuch as I have described in the preceding chapter the importance of *habit* for the subject of our analysis, patient nonadherence, it seems pertinent to describe it, or at least some aspects of it, in terms of *akrasia* in the *strict sense of the word*. However, after what has been said, it is clear that a change of preference related to the way we discount the future can *also* explain these choices, which then would only seem to be irrational.

Here we have arrived at a key in our understanding of patient nonadherence: The perception of time intervenes in the choice between an adherent and a nonadherent action in all cases. In certain cases, particularly where addiction is involved, the Theory of Intertemporal Choice accounts adequately for the role of time perception. But in most other cases, it is necessary to find another explanation for the intervention of time in what appears to be an incontinent action. The question 'why do we take care of ourselves?' might now be reformulated as: How is it possible for us to not conform to our own principle of continent action? How do I not take my pills, or rather, *how do I not make the decision to take my pills in the future?*

I propose that, in the particular case of *akrasia* represented by patient nonadherence to long-term therapies, there is such disequilibrium between the two types of actions, the continent and the incontinent, that it does not allow the principle of continence to express itself, or rather this principle becomes insufficient, or even inappropriate, if used alone. This leads me to propose a hypothesis where time plays a role in the partitioning of the mind proposed by Davidson in his explanation of irrationality; this hypothesis introduces a second principle, which I will call the *principle of foresight*, which pushes us to give priority to the propositional contents related to the future, i.e., to accept taking care of ourselves.

In short, then, the conflict between our reasons for continence and those for incontinence is well-matched and understandable. We find our mind at war with itself. It is only by recourse to another guiding principle, that of foresight, that we can see our way out of this dilemma.

7.2 The Principle of Foresight

Let's summarize. When we analyzed the taking of a pill as an action, it led us to consider patient nonadherence as a particular instance of incontinent action, or *akrasia*. As Davidson proposes in *Paradoxes of Irrationality*, incontinent action is caused by a motivation found in a part of the mind separated from the part containing the motivation that *could have* caused the continent action (which in our case would correspond to patient adherence). An akratic person's irrationality lies in defying the principle of continence that tells her to take into account all the available reasons, in whatever region of the mind they may be, before acting, and to act on the reason that she finds, all things considered, to be the best. This evaluation brings into play beliefs and desires, but also, as we saw with Livet, emotions. Signaling to the agent a differential between the reality of the world and her assessment of it, certain emotions, such as fear, indicate to her that a revision of preferences is necessary, while others, such as revulsion, push her to resist and maintain her position.

In this investigation, I defined patient adherence as 'to accept performing the repetitive actions recommended for a long-term health objective'. That is why patient nonadherence is most likely to occur in the case of a *chronic* illness, making it necessary to add the dimension of time to the analysis. Time may intervene negatively either by reducing the strength of the emotions that had pushed us at the beginning of the illness to take care of ourselves, or by leading to a change of preference at crucial turning points when nonadherence may occur, according to the mechanism described in the previous chapter. Boredom, the dull plodding of time, can deplete our ability to sustain adherence over the long haul. Energy for and interest in repetitive tasks wanes, and the resulting weariness can lead to giving up.

The desires and beliefs that drive the actions of adherence and nonadherence differ in their respective *relationship to time*: In a nutshell, the reward of adherence is far off, that of nonadherence is immediate. From this perspective, one might reasonably ask if the real question is not *how is adherence, rather than nonadherence, possible*, but rather: Why is anyone adherent? Why does one make the choice to take care of oneself?

7.2.1 Temporality as a Criterion for Sorting the Content of Mental States and the Principle of Foresight

The importance of time's role leads me to suggest the existence of a fundamental partitioning of the mind, wherein mental states are *sorted according to a temporality criterion*. In one of the areas of the mind are found the mental states that can be used for the motivations devoted to the present, in the other, the ones dedicated to the future, more precisely, *which are aimed to preserve the future*. If we say these two sub-domains of the mind exist, separated according to this temporal criterion, we can then say the adherent patient is the one who is capable of conforming to a *second principle*. Call it *the principle of foresight*: After the intervention of the principle of continence, it *advises her to give priority to the motivations oriented towards the future*, refusing instant rewards.

By itself, the principle of continence is impartial: It reflects the coherence of our rationality, but it doesn't tell us why we should pick one or the other of two options when there is a conflict. As Gjelsvik notes,

> Davidson's concept allows for causal/irrational deviations from what is considered best, all things considered. Theoretically, there is no reason to expect any particular pattern in these causal/irrational deviations. They might favor a long-term perspective or a short-term one. Davidson's theory gives no support to specific predictions and, in a sense, it therefore simply represents noise for a theory to attempt to predict people's behavior on the basis of their beliefs and desires (Gjelsvik 1999, 55).

But as we shall see, there is a profound imbalance between the attraction of future events and that of immediate events, a disequilibrium that tips the scale towards the latter. An additional principle is thus necessary to tip the scale towards the former. That additional principle must account for these different temporal horizons, and provide a mechanism whereby a person can escape the tyranny of immediate rewards. This additional principle is that of *foresight* (Reach 2008).

7.2.2 Implications of the Hypothesis

Imagine a person lacking this principle of foresight. When she considers all available information, she may very well conclude that her best choice is a sooner yet smaller reward. Those around her—her family, her physician—find her choice irrational, selfish, even outrageous; yet to her, the choice is perfectly rational. By just following the principle of continence, she chooses an action which everyone but she regards as irrational. We find ourselves in the awkward position of not having a way to decide whether a choice is rational or not. Or, perhaps even worse, we might conclude that "rationality is in the eye of the beholder", undercutting any attempt to bridge a profound divide between patient and physician.

Treatment nonadherence, then, is the result of absent, weakened or atrophied foresight. This would explain the phenomenon of nonadherence under the influence of alcohol (the effects of alcohol on unprotected sex are well known, and, less dramatically, aperitifs are served to encourage appetite) or nonadherence due to a temporary but overwhelming emotion (Gjelsvik 1999, 57). I know a former smoker, who started smoking again after five years of abstinence, the evening when he was eating dinner with his father who had just learned how serious his cancer was. Did he want by this gesture to go back to the way things were five years ago, when his family's happiness was intact? Peter Herman and Janet Polivy describe in detail the effects of September 11 on the abandonment of diets and alcohol and tobacco abstinence, as well as an excess of shopping, which is a compulsive behavior subject to control and to the risk of disinhibition (Herman and Polivy 2003, 478–479). We can also understand the dramatic effects of social deprivation on adherence (Wamala et al. 2007), inasmuch as social deprivation markedly diminishes a patient's ability to imagine her future. Social deprivation would inhibit the principle of foresight, or even completely destroy it.

On the other hand, we can understand how some patients, who were previously nonadherent to medical advice, suddenly become adherent thanks to an event laden with future-oriented connotations, such as the birth of a child or a grandchild.

We can also understand what we already mentioned: Adherence to medication (a bisphosphonate or a statin) was found in two studies to be associated with the use of preventive health services, such as prostate-specific antigen tests, fecal occult blood tests, screening mammography or influenza and pneumococcal vaccination (Curtis 2009; Brookhart et al. 2007). In the same vein, a study found an association between future preference, assessed by a four-item questionnaire, and the frequency of use of genetic counseling for BRCA1/2 testing (used to obtain information about future risk of breast and ovarian cancer) and mammography screening. The preference for future outcomes, determined through the four-item time-preference scale, was inversely correlated with monetary impulsivity and with cigarette smoking (Gurmankin Levy et al. 2006).

This is consistent with the idea that impatience, which can be evaluated through the monetary choice procedure, is a part of a more general construct, described as temporal horizon, which, according to Bickel et al. (2008) describes the window in time in which an individual is capable of perceiving and planning. In a recent study, temporal horizon was evaluated by using a method initially described by Wallace (1956), based on two types of questions: For instance, the first question asks participants to write a list of ten events that will occur in their lives and to indicate the age at which they would expect each event to occur. The second type of question asks participants to write endings to short stories and to indicate the duration of the story. Both measures provide estimates of how far into the future a person typically plans. Participants in the study also completed a monetary discounting task. Overall, participants with lower discounting rates tended to provide longer time estimates for stories than those with higher discounting rates. Smoking women were found to have a shorter temporal horizon than their non-smoking counterparts (Jones et al. 2009). This type of investigation has not, so far as I am aware, been performed in relation to adherence to medication or other health behaviors. However, this concept of temporal horizon may explain why people who live in poverty or any form of social deprivation and uncertainty, who have a short life expectancy, or who have other constraints upon their future expectations, may very reasonably opt for the short-term benefit and, therefore, be nonadherent to long-term therapies.

To substantiate this claim, I will show that the desires and beliefs leading to continent action are, by and large, oriented towards the future, whereas the ones leading to nonadherence are dedicated to the present.

7.2.2.1 Time and Desires

We can list the desires leading to a continent action such as taking a daily aspirin: "*I see something desirable in every action that* protects me from the risk of long-term complications, preserves my health, banishes the fear of the future, takes away the feeling of guilt that I would feel if I didn't do it, gives me a feeling of control", etc. And we can list the desires which lead to the incontinent action of not taking the aspirin: "*I see something desirable in every action that*: doesn't risk the undesirable side effects of the medication, doesn't take up my time, doesn't make me tired, and doesn't bore me." Clearly, the first group of desires is largely long-term oriented, while the second is short-term. We find once again the definition of opposing telic and paratelic states from Apter's Theory of reversion of mental states (see Chap. 2).

7.2.2.2 Time and Beliefs

Earlier in this book we found that general beliefs about health underpin our thinking about specific health issues as they arise. For example, beliefs that it is desirable to preserve one's health and that it is possible to do so, set the stage for seeking help when symptoms arise and for starting the prescribed treatment. Our general beliefs do not specifically dictate the types of efforts we will make—after all, the person who consults the psychiatrist and the person who consults the exorcist both seek a cure—but they do shape our decision to try (or not try). In general, an action is possible only if the person believes that the action will have meaningful results. Needless to say, if a person seeks to preserve her health, and to do so calls for an unpleasant or protracted treatment, she must have some belief that doing this treatment will be worthwhile in the end. This belief is necessarily future-oriented. For instance, one can't help but be struck by the analogy between health beliefs and religious beliefs

(Engel 1995, 68–94; Laplantine 1997, 382–388). First, health is existential in nature, as are religious beliefs. Second, as is the case with religious beliefs, individual health beliefs are also situated in a broader framework of collective beliefs. When Durkheim, discussing religious beliefs, argues for the necessity for a society to create an ideal, how can we fail to consider health as one of the ideals of modern society? Georges Canguilhem compares health to a lost paradise whence illness exiled us:

> Even when the disease becomes chronic, after having been critical, there is a past for which the patient or those around him remain nostalgic (Canguilhem 1989, 138).

and

> no cure is a return to biological innocence (Canguilhem 1989, 228).

Are we not dealing with an *ideal* and is this ideal not essentially collective? Canguilhem reminds us that the frontispiece of volume VI of the French Encyclopedia, 'The Human Being' published under the direction of the famous nineteenth century surgeon Leriche, represents health as an athlete, a shot putter (Canguilhem 1989, 196). To compare health to an athlete is it not to compare the exercise of health to sport, a collective activity?

Thirdly, and most importantly, just as the first article of faith in many religions is that they lead to salvation in the afterlife, health beliefs also concern the future—often, a distant, unseen future. In order to preserve one's future health it is necessary to believe today that one can act effectively. Thus, the beliefs that lead us to adherence, like the corresponding desires, are future-oriented. In the case of chronic illnesses such as diabetes, patients assemble into associations and self-help groups, they have meetings and news bulletins, offering patients real and immediate comfort when facing an uncertain future, not without parallels to the comfort sought by those who belong to a church.

Beyond these associations, there is what is called public health, which is not the health of the public, but health *for* the public. In other words, there is the ideal for which society needs to create, eventually through *consensus conferences*, collective beliefs. Can they be transmitted to individuals to serve as individual health beliefs? Perhaps the key to our problem is partly here: The differences between these collective beliefs and desires and individual mental states. The former define public health, and they are essentially future-oriented; whereas individual health beliefs and desires are more focused on quality of life, which quality is necessarily weighted towards the present and on the *hic* and *nunc* of our patients.

Jeremy Bentham, whose philosophy is at the origins of utilitarianism, has also noted the importance of the proximity of pleasure. Émile Durkheim, in a conference given in 1884, sums up the aspect of Bentham's thought consecrated to the arithmetic of pleasure, felicific calculus:

> Although there are many very different kinds of pleasures and pains, there are only a few dimensions on which they vary – intensity, duration, certainty, proximity. (Durkheim 2004, 234; Nordenfelt 1993, 17–34)

Public information campaigns, such as the No Tobacco Day, which comes back every year like a religious feast, aim to fill this gap. It is largely possible because some of the contents of the mental domain of the patient, the carrier of individual beliefs and desires, and that of her physician, transmitter of communal health beliefs and desires, are shared and can be publically expressed.

But even more importantly, this could be an essential aspect of medical practice: Through the sharing of desires and beliefs, to fill the mental domain dedicated to the patient's future, and to help her to form the principle of foresight that will lead her to take care of herself in the framework of a health *project*.

7.3 The Appearance of Adherence

> It is good to follow one's inclination as long as it rises.
> André Gide, The Counterfeiters

Adherence implies a desire to keep one's health, or, more generally, to not risk one's life. It seems to be related to our consciousness of existence, of time's passage and of our finitude. In this regard, it is doubtless an essentially human trait: Birds and animals generally choose the closest reward, whatever its real value. Nonetheless, some studies suggest that even pigeons are capable of choosing the more distant reward, although to a markedly lesser degree than humans. It is then legitimate to ask whether the capacity to wait, which makes adherence possible, is somehow physiological in nature, having developed according to the general laws of evolution.

As discussed by Patrick Pharo, this type of 'teleosemantic' question fits

> into the evolutionary tendencies of contemporary cognitive science, allowing us to finally rethink cognitive functions within the context of neo-Darwinism, by considering cognitive functions no longer as complex computational structures, but as the optimal result, although contingent on the evolutionary history of the species (Pharo 2006, 222).

7.3.1 From Animal to Human, a Phylogenesis of Patience

Time-related preferences are connected to an estimate of the value that a given good will have after a certain amount of time, say a year. This estimate takes into account both its real utility and the probability of its availability. If my environment is hostile, it may be sensible to consume a good sooner rather than later. Walter Mischel has observed that among Trinidadian children, those who lived in families where the father was absent were less capable of waiting for rewards (Mischel et al. 2003, 177). Thus, the preference for more immediate rewards could represent an innate evolutionary trait (Nozick 1993, 15, Footnote 17) developed in animals or early humans who could not engage in scholarly calculations of probability, in order to allow them to immediately enjoy the goods (food, for example) in a hostile environment which made their later availability uncertain.

As proposed by Elster (1977), "in the animal kingdom, the general mechanisms of natural selection create a presumption of shortsightedness". At the beginning of the last century, the French humorist Tristan Bernard was asked: If there were a fire at the Louvre Museum, which painting would you save? To which he responded: the one closest to the door. What is funny in this witticism is the animal, non human aspect of the answer.

But evolution may also have selected mechanisms leading animals to act with some degree of patience. Squirrels make reserves of food for the winter and birds do not lay their eggs just any time: They do it when the nutritional needs of newly hatched young birds and an abundance of caterpillars are synchronized, while impatient parents would put at risk the future of their offspring. There may be a real evolution over time of the *function of patience* (Kacelnik 2003, 116) living beings became little by little capable of resisting the idea of *a bird in the hand is worth two in the bush*. We might be tempted to think that in the case of animals, patience is innate, hard-wired behavior, while in humans, delay of gratification is purely learned. In reality, patience in animals and humans have biological determinants which shape the adaptivity of delayed gratification.

7.3.2 Development of Patience in Children: Ontogenesis

According to a biological paradigm, 'ontogeny recapitulates phylogeny'. That is, during our embryonic and fetal development and during infancy, we repeat the steps of our distant ancestors on the road towards our species. The same may be true for the function of patience. Walter Mischel's studies, using his famous marshmallow test, show the gradual appearance of the ability to postpone obtaining a reward during infancy, with the progressive dominance of what he calls the 'cool', cognitive, regulatory system, which allows the realization of long-term projects, over the 'hot' system, which is present at birth, emotional, impulsive, and which allows one to engage in immediate consumption or evasive reactions (fight or flight) (Mischel et al. 2003, 175–200). Mischel's 'cool-hot' model, which is not without similarities with the two models we saw in the first part of this book, those proposed by Leventhal and Apter, stipulates that the two systems interact. For example, children were capable of delaying acquiring sweets, when they had a choice, by appealing to their cool system: They were asked to imagine its most abstract aspects—for instance, a marshmallow looks like a big white cloud. On the other hand, when they were asked to imagine the sweet and melt-in-your-mouth aspect of the marshmallow, they could not wait longer than 5 min. Before the age of four, this capacity is absent and is not completely developed until the age of 12. It is interesting to note that the four year old children who were most capable of waiting had obvious intellectual advantages in terms of their ability to plan projects and had superior academic performance as adolescents (Mischel et al. 1989).

7.3.3 Neuroanatomy of Patience

We are beginning to understand the neurophysiology underpinning the function of patience. Antonio Damasio's famous book, *Descartes'* Error, (Damasio 1994) popularized the story of Phineas Gage who, in 1848, survived a terrible lesion to his prefrontal cortex. He survived, but his personality changed. He became incapable of making decisions or of showing socially appropriate emotions. Studies conducted by Damasio's team with patients suffering from lesions to this area of the brain suggest that it plays a part in our capacity to avoid immediate rewards if they have negative consequences for us.

It is the same for rats. This is important for this discussion because it suggests that the human patience may piggyback on a function already present in animals: Experimental lesions of the orbitofrontal cortex or the septum (the region which links the limbic system to the orbitofrontal cortex) shorten the period an animal would normally tolerate when presented with a choice between an immediate reward and a larger but later reward. So perhaps this neural area underwrites the function of patience. And since the human frontal cortex is relatively outsized, this may explain our relatively deep reserves of patience (Manuck et al. 2003, 144–148).

These neuroanotomical hypotheses are in agreement with experimental studies. For example, they mesh with Joseph LeDoux's attempts to locate the cerebral area of emotions (LeDoux 1996). And Mischel suggests that the amygdala is the cerebral location of the hot, emotional, impulsive system, which allows us to respond to danger immediately, and which is present in the infant from birth. Meanwhile, he hypothesizes that the cool, cognitive system is located in the hippocampus and prefrontal cortex. These two structures develop around the age of 4, which is also the time when children acquire the capacity to delay obtaining a reward (Mischel et al. 2003, 180).

Schematically, when faced with a perceived danger, a quick but imprecise response is possible. Information is transmitted by the thalamus from the visual cortex to the amygdala, setting off the emotional response (for example, immediate flight) to the impression that, say, the form we see on the floor is a snake. A more elaborate, slower, reaction passes through the high road of the cortex and the hippocampus: "No, after all it's only a piece of wood." Studies using brain functional MRI show that the degree of impulsivity manifested during a monetary choice can be seen as the result of the activity of two competing systems: Choices for delayed outcomes are related to the prefrontal cortex, while choices for immediate outcomes are related to the limbic brain regions (Bickel et al. 2007). A recent study in teenagers and young adults demonstrated an association between the reduction in impulsivity observed in post-pubertal years and the maturation of limbic frontostriatal circuitry (Christakou et al. 2011).

7.3.4 Neurobiology of Patience

Nerve cells communicate with each other through neurotransmitters, chemical substances released in the synaptic cleft separating them. The nature of neurotransmitters involved in the biochemistry of patience is beginning to be more precisely understood. Serotonin appears to play a major role: A study demonstrated a strong correlation between patience, as determined by a questionnaire, and the levels of secretion of cerebral serotonin, evaluated by measuring the level of the pituitary hormone prolactin stimulated by fenfluramine (Manuck et al. 2003, 156). Individuals who have a weak secretion are those who appear more impulsive, saying they prefer the present to the future, spend more than they earn, are not economical, make few plans for the future, and, finally, are negligent concerning their medical and dental checkups. This of course speaks directly to the subject of this book.

One can now imagine the potential pharmaceutical implications of these data: Currently we have at our disposal numerous medications capable of increasing cerebral serotonin, for instance by inhibiting its reuptake once it has been released into the synaptic cleft, or its breakdown by an enzyme, monoamine oxidase (MAO), and finally by stimulating its secretion, like with fenfluramine. This could lead to a real pharmacology of patience (Manuck et al. 2003, 158–162). The Selective Serotonin Reuptake Inhibitors (SSRIs) seem typically to have an effect on patience: For example, many patients report that they feel less "caught up" by interpersonal irritants; they feel they can let things go and focus more on the big picture. They also report less irritability (the opposite of patience?) and greater sense of ease (important for delaying gratification).[2] An effect of SSRIs on delaying gratification was indeed demonstrated in pigeons (Wolff and Leander 2002).

7.3.5 Genetics of Patience

Patience may have a partly genetic basis: Normally, serotonin is broken down by MAO. The stronger the enzyme's activity, the lower the levels of cerebral serotonin. There is a polymorphism in the gene coding for this enzyme: Individuals who have inherited certain forms of the gene—or alleles—will express it more strongly, produce more MAO and consequently have lower levels of cerebral serotonin (for example, their level of circulating prolactin does not increase as much after stimulation with fenfluramine). Stephen Manuck's group has similarly shown that these individuals had a higher impulsiveness score (Manuck et al. 2003, 158). However, there was an important overlap between values observed with individuals who had these alleles and others, suggesting that genetics, as usual, does not explain everything and that the personalities formed as a result of one's environment play an important role.

In sum, we see the emergence of a biological conception of patience (Kalenscher and Pennartz 2008), and this may be relevant for the study of patient adherence to long-term therapies (Reach 2010). The field studying this most rigorously, neuro-economics, is expanding quickly. A study showed that the degree of impatience, evaluated by a monetary choice, was significantly more

---

[2] I am grateful to John Meyers for this remark.

pronounced in alcoholic individuals, totally abstinent for at least one year, than in non-alcoholic controls. Brain functional MRI evaluation of these individuals made during the monetary choice task demonstrated significant correlations between their levels of impulsivity, determined by their choice, and the function of certain zones of the brain known to be active during decision-making processes. For example, they found a significant negative correlation with the activity of the orbitofrontal cortex that is supposed to influence the capacity to wait for long-term rewards. The same study showed that there was a relationship between a genetic polymorphism of the catechol-O-methyltransferase (which intervenes in the metabolism of the dopamine) and a person's degree of impulsivity on the one hand, and the functional activity of these cerebral zones on the other (Boettiger et al. 2007, see also Kalis et al. 2008).

All this could help explain the observation that there are patient people and impatient people, or, to use Elster's expression, "certain [people] like the present, while others have a taste for the future". It could also explain how emotions, stress or alcohol can tip us over into impatience through a purely neurobiological action affecting one of the cerebral areas. These cerebral functions may have evolved from similar structures in animals which trigger us to flee in the face of danger. Little by little, thanks to the enormous development of the frontal cortex, humans have become capable of balancing this quick, emotional system with a more reflective system of cognition.

The emotional system clearly has a selective advantage as it allows flight: In the ancestral environment, a quick escape would allow one to live—and propagate—another day. So it is not surprising that it developed under evolutionary pressure. But the same could be true for the cognitive system which allows us to coolly calculate that it is advantageous to be economic: Believing that it is good to keep some food in provision for periods of scarcity, rather than eating it immediately, could have been a selective advantage for the first humans (this is the meaning of Pharaoh's dream, as interpreted by Joseph). This advantage would have been particularly powerful: On the one hand, it did not depend on the slow and random selection through mutations; on the other hand, it was available *to all of us*, while the biological capacity of storing nutrients in the form of fat in provision of scarcity, which we mentioned earlier for Pima Indians, was only transmitted through the genetic lottery, to *certain lucky individuals*.

7.3.6 The Appearance of Belief

In his essay devoted to the nature of rationality, Nozick (1993, 93) asks a simple question: *Why do we have beliefs?* He suggests that organisms who act on the basis of beliefs have an advantage over those who only have hardwired behaviors producing automatic responses to stimuli. Of course, the stimulus-response formula can, as behaviorists argue, have a certain degree of adaptability thanks to the mechanisms of conditioning (according to the operant conditioning theory developed by B.F. Skinner, the consequences of a behavior reinforce or decrease the probability that the behavior will be performed again in similar circumstances) but it is a slow process incapable of quickly adapting to new situations, produces maladjusted or obsolete behavior, and requires sustained reinforcement to develop (Nozick 1993, 93–94).

But here the stimuli are in fact the state of the world around us, a world that's constantly and unexpectedly changing. Faced with this situation, evolution's answer was double: First, the selection of mechanisms producing rational representation in the brain of the state of the world in the form of a mental puzzle, whose pieces can be put together and taken apart all the while maintaining a certain coherence, taking into account our desires, i.e., our projects, and our emotions, which allows the more primitive system to take the helm in case of danger; second, the development of a rationality of action which puts it under the control of desires, beliefs, and emotions. Since it is sometimes dangerous to check whether the suspicious object is indeed a snake, maybe it is better for one's fear and repulsion to make one flee. But some individuals come to refine their knowledge of snakes and become capable of resolutely following on their way (Nozick 1993, 113).

Ramsey showed that the degree of a belief's truth is equal to the probability that this belief will lead to a successful action. What evolution must have selected were the processes which allow us to keep only beliefs whose content is relatively close to the truth (they do not need to be entirely true, and this is the advantage of acting on belief and not knowledge). Otherwise, the result of our actions would be random. But if we follow this evolutionary argument developed by Nozick, we still need to resolve two problems. On the one hand, beliefs must be consulted when deciding to act, but the mind's complexity may impede such consultation. On the other hand, since actions must be compatible with the species' survival on the whole, desires leading to more evolutionarily advantageous actions must be favored.

To overcome these difficulties, natural selection, or, as we shall see, a new type of selection relying on inter-human communication, retained a particularly efficient method for the development of rationality: Principles.

In the course of our analysis, we have encountered three "principles": (1) What Davidson called the principle of warrant, telling us to take into account all the pertinent arguments before choosing one; failure to use it leads to self-deception, and in medicine, to the denial of illness. (2) The principle of continence, proposed by Davidson, which tells us to perform the action which, all things considered, we think is the best; flouting it leads to incontinent actions. (3) The principle of foresight, proposed in this book, which leads us to give priority to actions inclined to guarantee our future; ignoring it leads to nonadherence.

The very notion of principle could be criticized if we used principles on an as-needed basis. One could then have to wonder why in certain cases the principle is not applied, obliging to admit the need for imagining the existence of a new principle enjoining to use the preceding principle, and one would thus be likely to arrive at an infinite regression.

But this criticism fails if we consider the principle to be a real device (to use Nozick's term) with a function; that is, if we imagine a principle as a tool. For instance, the function of the principle of warrant is to impel us consider all the available arguments. This means seeing in principles the rationality's way of surmounting its own difficulty: The absence of laws, a consequence of the holistic character of the mind. If there are no laws, says Elster, we must content ourselves with 'mechanisms', which allow us, after the fact, to explain the observed behavior. But after the fact is perhaps too late, and the absence of laws could result in haphazard, *au petit bonheur*, reasoning or decision making.

Principles could fill the absence of laws in the brain: As Nozick says,

> we are creatures who do not act automatically, without any guidance. [...] Doesn't this show that the purpose of principles is to guide us to something, whatever that is, that we would not reach by acting at random? (Nozick 1993, 40)

Otherwise, without laws, we would roll with the slope, moved only by laws of physics which, in their universality, apply even to us. According to a metaphor taken from Kacelnik (2003, 116), a river under its influence will flow down more and more rapidly, following the steepest hill, and will never spontaneously go up the hill, even if it were to find a steeper valley as a result. *It is good to follow one's inclination as long as it rises,* says Gide. This is what we do by giving priority to the future; *it is good,* but this goes against a powerful natural tendency. Such a turn away from the path of least resistance becomes possible thanks to the development of principles.

We can then say that principles have an advantage, from the evolutionary point of view. They offer to the one who honors them an advantage in terms of survival: Like the crafty, and ingenious Ulysses, he who lives in accord with principles outlives those less so inclined.

7.4 A Pathophysiological Point of View

These pathophysiological metaphors—the failure of a principle to explain denial, incontinent actions, patient nonadherence—may seem naïve, and they certainly reflect my experience as a clinician trained in a deeply pathophysiological context (they are well adapted to diabetology, endocrinology...). But this investigation is reminiscent of the path that for Canguilhem leads from pathology to physiology:

> Health is organic innocence. It must be lost, like all innocence, so that knowledge may be possible. Physiology is like all science, which, as Aristotle says, proceeds from wonder. But the truly vital wonder is the anguish caused by disease (Canguilhem 1989, 101).

And so it is to explain how some patients *can* be nonadherent and others *can* be adherent that I was led to postulate the existence of the somewhat physiological principle of foresight. This is similar Davidson's postulate of the principle of continence to explain weakness of the will and the principle of warrant to explain self-deception, these two principles intervening 'generally', one is tempted to say 'physiologically', i.e. 'normally'.

This is exactly the pathophysiological path described by Georges Canguilhem:

The physician has a tendency to forget that it is the patients who call him. The physiologist has a tendency to forget that a clinical and therapeutic medicine, which was not always so absurd as one might think, preceded physiology. Once this oversight is remedied, we are led to think that it is the experience of an obstacle, first lived by a concrete man in the form of disease, which has given rise to pathology in its two aspects, clinical semiology and the physiological interpretation of symptoms. If there were no pathological obstacles there would be no physiology because there would be no physiological problems to solve. [...] we can say that in biology it is the *pathos* which conditions the *logos* because it gives it its name. It is the abnormal which arouses theoretical interest in the normal. Norms are recognized as such only when they are broken. Functions are revealed only when they fail. Life rises to the consciousness and science of itself only through maladaptation, failure and pain (Canguilhem 1989, 208–209).

In biology, says Canguilhem. Why would it not be the same for the mind? By proposing the existence of the principle of continence, Donald Davidson did not simply posit an ad hoc explanation to allow a way out of the dead end of *akrasia*: He observed a surprising phenomenon, which led him to describe the missing function. Canguilhem reminds us that we had to wait for Addison's clinical description of the insufficiency of the adrenal gland to understand its function (a century earlier, the jury of a competition to discover its function, presided by Montesquieu in Bordeaux, had to give up on awarding the prize): The function of the adrenal gland is to avoid adrenal insufficiency, whose symptoms were described by Addison. Would it be justified to say that this explanation was only an ad hoc explanation? A century after Addison's description, Kendall, working at the Mayo Clinic, isolated the hormone, cortisone, which is now used in the treatment of this disease.

In this analysis, to explain patient nonadherence we have proposed a role for time in the partitioning of the mind as well as the need to give priority to the future (principle of foresight). This notion of arrangement according to mental criteria can also seem naïve; it is not far from the one proposed by Damasio in a neurophysiological, not philosophical, context, to explain how the brain functions:

Whether we conceive of reason as based on automated selection, or on a logical deduction mediated by a symbolic system, or – preferably – both, we cannot ignore the problem of order. I propose the following solution: (1) If order is to be created among available possibilities, then they must be ranked. (2) If they are to be ranked, then criteria are needed (values or preferences are equivalent terms). (3) Criteria are provided by somatic markers, which express, at any given time, the cumulative preferences we have both received and acquired (Damasio 1994, 199).

One may wonder if we don't have here something that is essentially human. Non-human animals, generally, do not care about time, and gods—or nature—are immortal. Perhaps organized propositional attitudes within a generally coherent holism is a peculiarly human trait. And perhaps the same can be said of amassing in one domain of the mind the contents of the propositional attitudes pertaining to the future; and, finally, the principle of foresight orients our choices by imposing a preference for the arguments located in the domain devoted to the future. One can also speculate that this differentiation is accomplished slowly in adults, leading from the simple *age of reason* to an *age of foresight*. This could explain the maturational pattern seen in many patients who come to adherence: A grasp of the necessity of their treatment, not merely on cognitive grounds but on a deeper emotional/moral plane. More mundanely, perhaps the transition from reason to foresight coincides with the decision to buy a life insurance policy, or to take care of oneself for the sake of others.

Indeed, the same could be true for the principle of continence. As Ruwen Ogien notes in his analysis of the weakness of the will,

> by evoking Saint Augustine's famous prayer, "give me chastity and continence, but not yet", Davidson seems to suggest that accepting the principle of total information is a virtue that is acquired slowly and with difficulty rather than a natural tendency (a psychological rule) or a law of logic which one rapidly comes to know. The one who has not yet acquired the principle of continence is not irrational (besides, he needs rationality to acquire continence) (Ogien 1993, 83).

We cannot but think about the Freudian distinction between the pleasure principle and the reality principle:

> We know that the pleasure principle is proper to a primary method of working on the part of the mental apparatus, but that, from the point of view of the self-preservation of the organism among the difficulties of the external world, it is from the very outset inefficient and even highly dangerous. Under the influence of the ego's instincts of self-preservation, the pleasure principle is replaced by the reality principle. This latter principle does not abandon the intention of ultimately obtaining pleasure, but it nevertheless demands and carries into effect the postponement of satisfaction, the abandonment of a number of possibilities of gaining satisfaction and the temporary toleration of unpleasure as a step on the long indirect road to pleasure (Freud 1990, 6).

These considerations may shed light on nonadherence observed in the adolescent. For instance, a study found direct evidence of poor compliance with insulin therapy in young patients with type 1 diabetes. The authors suggested that poor adherence to insulin treatment was the major factor that contributes to long-term poor glycemic control and diabetic ketoacidosis in this age group (Morris et al. 1997), an age when one often lives in the present, although other factors certainly play a role as well.

This principle of foresight seems to appear well after the simple capacity to delay gratification. And since the latter appears progressively around age 4, with 60 % of the children having acquired the maximum waiting capacity used in the test (20 min) at age 12 (Mischel et al. 2003, 187), it would be unsurprising if the principle of foresight does not appear until after adolescence.

7.5   A Top-down Model of Adherence

The principle of foresight is analogous to other concepts we have encountered in this study: Self control, will-power, or strength of the will.

There is therefore a difference between the conception of adherence presented here, where the patient is adherent if she applies rational principles (continence and foresight), and that proposed by Ainslie, where the patient applies her own 'personal rules' to each of the intertemporal choices she faces. In the Ainslie model, it is not necessary to appeal to an authority called 'the will'. Personal rules are a *tactic,* which bundles all the individual choices into a single decision, say, to never smoke again; it is a 'molar' choice. On the contrary, appealing to principles of continence and foresight means developing a global *strategy*, which can be called 'supramolar', wherein the patient brings together all the different aspects of adherence. Not only does she stop smoking, but she also takes her medications, stays on her diet, etc., in a global strategy designed to preserve her future. She may ask herself: Wouldn't it be stupid to lose all the benefits of my efforts to stop smoking by not following a diet? As Holton says, it then becomes possible for her

> to abide by all of one's resolutions: resolutions not to drink, not to smoke, to eat well, to exercise, to work hard, not to watch daytime television, or whatever (Holton 2003).

We may ask why, when an agent is adherent because she has the two principles of continence and foresight, she is not adherent for all the aspects of her therapy. In reality, this situation is pretty rare: We have mentioned at the beginning of the book that only 7 % of diabetic patients are adherent to all medical advice they receive. And yet, as we have also noted, nonadherence or adherence form a whole, in accordance with the hypothesis that they are homologous phenomena with the same mechanism (in medicine, we would call it a syndrome): I have quoted at the beginning of this book Horwitz's observation that not being adherent to a placebo raises the risk of death (Horwitz et al. 1990). This can be explained if we admit that people not adherent in taking their pills (here, a placebo), are also nonadherent when it comes to other recommendations which are supposed to protect them. I have also quoted data from literature showing that smokers are less adherent to medical recommendations than others. The same holds true for alcohol consumption (Ahmed et al. 2006). Conversely, we can show that patients who often measure their glycemia are also those who devote more time to their diet, exercise, etc. (Safford et al. 2005).

These rational 'supramolar' principles intervene at the tactical level of personal rules which push the individual to weigh the choice between an immediate and a delayed reward, as a test allowing her to predict her future choices.

Figure 7.5 represents this conception of adherence (Reach 2008). According to this graphic, nonadherence—or weakness of the will—can have two descriptions, which, in turn, can explain adherence (Fig. 7.5). In the first, proposed by George Ainslie, the agent succumbs to a reversal of preferences, which occurs at the time of 'intertemporal choice'. She can fight against the risk of 'caving in' and become adherent by forging a personal rule for a given problem, represented by the little figures borrowed from Ainslie (see Fig. 7.4). The other description of weakness of the will is based on the failure of the rational principles of continence and of foresight. These principles influence different aspects of adherence, and they could also represent a force which leads the agent to adopt personal rules concerning particular aspects of adherence. In this second role, the principles confer symbolic value on Ainslie's personal rules; without this value, they would have a purely behaviorist significance.

Ainslie's tactic of personal rules also takes into account the future, but, whether the agent puts it into effect under the influence of a behaviorist reinforcement or

Fig. 7.5 Adherence, two
possible descriptions.
Originally published in
(Reach 2008), © Reach 2008

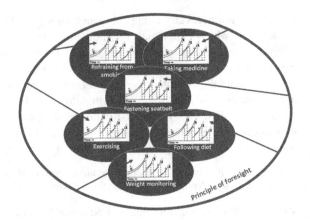

consciously, it remains linked to one type of behavior. It can even be limited to an activity within a certain context (a personal example: I managed to get into the habit of never, without exception, taking the elevator when I'm at the hospital, but not when I'm home).

On the other hand, systematic use of the principle of foresight both integrates temporality (something the principle of continence fails to do) and supposes that we do not content ourselves with considering an action in its immediacy. Moreover, it has a global effect: It reflectively concerns all the individual's activities. It is in this global context that an individual *wants to want* and exercises her capacity to have second order desires. Indeed, it is this very quality which distinguishes the human from other creatures, according to Frankfurt (1988). As Joëlle Proust remarks, this model

> concerns the large *orientations* – principles or motivations – which *determine the choice of pertinent actions in the middle and long-term,* such as social, artistic and emotional objectives, utilities and values which form the agent's preferences (Proust 2005, 294–295).

Thus, we arrive at a 'top-down' model of adherence, *supposing, finally, the existence of a will,* which allows the use of this strategy. This model proposes that when we speak of "weakness of the will" (*akrasia*) or "strength of the will" (*enkrateia, willpower*), there is truly a 'faculty', the will, which, depending on the case, is weak or strong, and which has the possibility, through the use of corresponding volitions, to influence all of the agent's projects: *The will exists, since it has its own pathology.* It finally appears that our research, originating as a purely medical problem (why do we take care of ourselves?) permits two wider applications: It allows us to propose a quasi-physiological explanation of patient adherence from the observation of nonadherence, and it can contribute to the purely philosophical debate concerning the nature—and the very existence—of the will.

7.5.1 Transmission of Principles

Robert Nozick insists on the symbolism of principles: They aim to give a symbolic value to an action by charging it with the weight of a whole category of actions.

But there is more: Their use itself has a symbolic value, making rationality not simply 'instrumental'. It makes actions efficient, and additionally:

> it is symbolically important to us that not all of our activities are aimed at satisfying our given desires. [...] One way we are not simply instrumentally rational is in caring about symbolic meanings, apart from what they cause or produce (Nozick 1993, 138–139).

We may now ask how a symbolic activity can be transmitted from generation to generation. In a purely biological, reductionist perspective we could certainly imagine that even the most symbolic thought has a neurophysiological substratum which is transmitted like all the other functions of the organism through genes.

But there is another possibility. Interpersonal communication allows a different mode of transmission. Nozick (1993, 126) mentions the possibility that rationality has a function of transmitting information between institutions. In this case, the principles could be *memes*, the new type of replicator proposed by Richard Dawkins in the *Selfish Gene*:

> just as genes propagate themselves in the gene pool by leaping from body to body via sperms and eggs, so memes propagate themselves in the meme pool by leaping from brain to brain via a process which, in the broad sense, can be called imitation. If a scientist hears, or reads about, a good idea, he passes it on to his colleagues and students. He mentions it in his articles and his lectures. If the idea catches on, it can be said to propagate itself, spreading from brain to brain (Dawkins 1976, 192)

To illustrate the concept of meme, Daniel Dennett, in *Consciousness Explained,* lists:

> the sort of complex ideas that form themselves into distinct memorable units – such as the ideas of : wheel, wearing clothes, vendetta, right triangle, alphabet, calendar, *the Odyssey*, calculus, chess, perspective drawing, evolution by natural selection, Impressionism, 'Greensleeves', deconstructionism (Dennett 1992, 201, italics mine).

7.5.2 Medicine and Health

Coming back to the question which was the subject of our book, *why do we take care of ourselves?*, this evolutionist interpretation finally suggests that being capable of taking care of oneself in order to protect one's health represents the prolongation of a vital instinct. This was Canguilhem's profound hypothesis:

> It is true that in medicine the normal state of the human body is the state one wants to reestablish. But is it because therapeutics aims at this state as a good goal to obtain that it is called normal, or is it because the interested party, that is, the sick man, considers it normal that therapeutics aim at? We hold the second statement to be true [...] We think that in doing this the living human being, in a more or less lucid way, extends a spontaneous effort, peculiar to life, to struggle against that which obstructs its preservation and development taken as norms. [...] We ask ourselves how a human need for therapeutics would have engendered a medicine which is increasingly clairvoyant with regard to the conditions of disease if life's struggle against the innumerable dangers threatening it were not a permanent and essential vital need (Canguilhem 1989, 126–127).

And so medicine was born, no longer having as its only goal to cure acute diseases, but also to prolong our future and increase our well-being. It was first a question of preventing complications of chronic diseases, then of preventing the very appearance of disease. Now, in Public Health, it is even the risks that we want to avoid, the principle of foresight of each one of us aroused by the campaigns resembling a real collective protection program, exhorting us to eat well, to exercise, to not smoke, to buckle our seatbelts. Following this evolution, soon will come a time when we will want to protect ourselves from even hypothetical risks—not unlike the shift in environmentalism from simply cleaning up the environment to preserving it. The individual principle of foresight and this collective precautionary principle display the same concern for the future.

Medicine is intimately involved in the vital momentum of life itself, as described by Bergson (2007, 295) in *Creative Evolution*:

> As the smallest grain of dust is bound up with our entire solar system, drawn along with it in that undivided movement of descent which is materiality itself, so all organized beings, from the humblest to the highest, from the first origins of life to the time in which we are, and in all places as in all times, do but evidence a single impulsion, the inverse of the movement of matter, and in itself indivisible. All the living hold together, all yield to the same tremendous push. The animal takes its stand on the plant, man bestrides animality, and the whole of humanity, in space and in time, is one immense army galloping beside and before and behind each of us in an overwhelming charge able to beat down every resistance and clear the most formidable obstacles, perhaps even death.

References

Ahmed AT, Karter AJ, Liu J. Alcohol consumption is inversely associated with adherence to diabetes self-care behaviours. Diabet Med. 2006;23:795–802.

Ainslie G. Beyond microeconomics, conflict among interests in multiple self as a determinant of value. In: Elster J, editor. The multiple self. Cambridge: Cambridge University Press; 1985.

Ainslie G. Picoeconomics: the strategic interaction of successive motivational states within the Person. Cambridge: Cambridge University Press; 1992.

Ainslie G. A selectionist model of the ego: implications for self-control. In: Sebanz N, Prinz W, editors. Disorders of volition. Cambridge: MIT Press; 2006. www.picoeconomics.com/Articles/MunichRepr.pdf, 2003.

Basil MD, Basil DZ, Schooler C. Cigarette advertising to counter New Year's resolutions. J Health Commun. 2000;5:161–74.

Baumeister R, Voes K. Willpower, choice and self-control. In: Loeweinstein G, Read D, Baumeister RF, editors. Time and decision, economic and psychological perspectives on intertemporal choice. New York: Russell Sage Foundation; 2003.

Bergson H. Creative evolution. Basingstoke: Palgrave Macmillan; 2007.

Bickel J, Johnson MW. Delay discounting, a fundamental behavioral process of drug dependence. In: Loewenstein G, Reid D, Baumeister RF, editors. Time and decision. New York: Russell Sage Foundation; 2003.

Bickel WK, Miller ML, Yi R, Kowal BP, Lindquist DM, Pitcock JA. Behavioral and neuroeconomics of drug addiction: competing neural systems and temporal discounting processes. Drug Alcohol Depend. 2007;90S:S85–91.

Bickel WK, Yi R, Kowal BP, Gatchalian KM. Cigarette smokers discount past and future rewards symmetrically and more than controls: is discounting a measure of impulsivity? Drug Alcohol Depend. 2008;96:256–62.

Boettiger CA, Mitchell JM, Tavares VC, Robertson M, Joslyn G, D'Esposito M, Fields HL. Immediate reward bias in humans: fronto-parietal networks and a role for the catechol-O-methyltransferase 158$^{val/val}$ genotype. J Neurosci. 2007;27:14383–91.

Brookhart MA, Patrick AR, Dormuth C, Avorn J, Shrank W, Cadarette SM, Solomon DH. Adherence to lipid-lowering therapy and the use of preventive health services: an investigation of the healthy user effect. Am J Epidemiol. 2007;166:348–54.

Canguilhem G. The normal and the pathological (trans: Fawcett C). New York: Zone Books; 1989.

Chapman GB. Time discounting of health outcomes. In: Loewenstein G, Read D, Baumeister RF, editors. Time and decision. New York: Russel Sage Foundation; 2003.

Chapman GB, Elstein AS. Valuing the future: temporal discounting of health and money. Med Decis Making. 1995;15:373–86.

Christakou A, Brammer M, Rubia K. Maturation of limbic corticostriatal activation and connectivity associated with developmental changes in temporal discounting. Neuroimage. 2011;54:1344–54.

Curtis JR, Xi J, Westfall AO, Cheng H, Lyles K, Saag KG, Delzell E. Improving the prediction of medication compliance: the example of bisphosphonates for osteoporosis. Med Care. 2009;47:334–341.

Damasio AR. Descartes' error. Emotions, reason and the human brain. New York: G.P. Putnam's Sons; 1994.

Dawkins R. The selfish gene. Oxford: Oxford University Press; 1976.

Dennett D. Consciousness explained. New York: Back Bay Books; 1992.

Durkheim E. Philosophy lectures, notes from the Lycée de Sens, 1883–1884 (trans: Gross N, Jones RA). Cambridge: Cambridge University Press; 2004.

Elster J. Ulysses and the sirens: a theory of imperfect rationality. Soc Sci Inf. 1977;16:469–526.

Elster J. In: Elster J, Skog O-J, editors. Getting hooked, rationality and addictions. Cambridge: Cambridge University Press; 1999a.

Elster J. Davidson on weakness of the will and self deception. In: Hahn LE, editor. The philosophy of Donald Davidson. Open Court; 1999b.

Elster J. Strong feelings, emotion, addiction and human behavior. The 1997 Jean Nicod lectures. A Bradford book. Cambridge: The MIT Press; 2000a.

Elster J. Ulysses unbound. Cambridge: Cambridge University Press; 2000b.

Engel P. Les croyances. In: Notions de philosophie (under the direction of D. Kamboucher). Paris: Editions folio, Gallimard; 1995.

Frankfurt H. The importance of what we care about. Cambridge: Cambridge University Press; 1988.

Frederick S. Time preference and personal identity. In: Loewenstein G, Read D, Baumeister RF, editors. Time and decision. New York: Russel Sage Foundation; 2003.

Freud S. Beyond the pleasure principle. New York: Norton Library; 1990.

Gjelsvik O. Addiction, weakness of the will and relapse. In: Elster J, Skog O-J, editors. Getting hooked, rationality and addiction. Cambridge: Cambridge University Press; 1999.

Gurmankin Levy A, Micco E, Putt M, Armstong K. Value for the future and breast cancer-preventive health behaviour. Cancer Epidemiol Biomark Prev. 2006;15:955–60.

Herman CP, Polivy J. Dieting as an exercise in behavioral economy. In: Loewenstein G, Read D, Baumeister RF, editors. Time and decision. New York: Russell Sage Foundation; 2003.

Homer. The Odyssey. Penguin; 2003.

Holton R. How is strength of will possible? In: Stroud S, Tappolet C, editors. Weakness of will and practical irrationality. Oxford: Clarendon Press; 2003.

Horwitz RI, Viscoli CM, Berkman L, Donaldson RM, Horwitz SM, Murray CJ, Ransohoff DF, Sindelar J. Treatment adherence and risk of death after myocardial infarction. Lancet. 1990;336:542–5.

Jones BA, Landes RD, Yi R, Bickel WK. Temporal horizon: modulation by smoking status and gender. Drug Alcohol Depend. 2009;104(Suppl 1):S87–93.

Kacelnik A. The evolution of patience. In: Loewinstein G, Read D, Baumeister RF, editors. Time and decision. New York: Russel Sage Foundation; 2003.

Kalenscher T, Pennartz CMA. Is a bird in the hand worth two in the future? The neuroeconomics of intertemporal decision-making. Prog Neurobiol. 2008;84:284–315.

Kalis A1, Mojzisch A, Schweizer TS, Kaiser S. Weakness of will, akrasia, and the neuropsy-chiatry of decision making: an interdisciplinary perspective. Cogn Affect Behav Neurosci. 2008;8:402–17.

Laplantine F. Anthropologie de la maladie. Bibliothèque Scientifique Payot; 1997.

LeDoux J. The emotional brain. New York: Touchstone; 1996.

Livet P. Émotions et rationalité morale, P. U. F., Collection Sociologies; 2002.

Manuck SB, Flory JD, Muldoon MF, Ferrel RE. A neurobiology of intertemporal choice. In: Loewenstein G, Read D, Baumeister RF, editors. Time and decision. New York: Russell Sage Foundation; 2003.

Mischel W, Shoda Y, Rodriguez ML. Delay of gratification in children. Science. 1989;244:933–8.

Mischel W, Ayduck O, Mendoza-Denton R. Sustaining delay of gratification over time: a hot-cool systems perspective. In: Loewenstein G, Read D, Baumeister RF, editors. Time and decision. New York: Russell Sage Foundation; 2003.

Morris AD, Boyle DI, McMahon AD, Greene SA, MacDonald TM, Newton RW. Adherence to insulin treatment, glycaemic control, and ketoacidosis in insulin-dependent diabetes mel-litus. The DARTS/MEMO collaboration. Diabetes audit and research in Tayside Scotland. Medicines Monitoring Unit. Lancet. 1997;350:1505–10.

Norcross JC, Mrykalo MS, Blagys MD. Auld lang syne: success predictors, change processes, and self-reported outcomes of New Year's resolves and nonresolvers. J Clin Psychol. 2002;58:397–405.

Nordenfelt L. Quality of life, health and happiness. Avebury; 1993.

Nozick R. The nature of rationality. Princeton: Princeton University Press; 1993.

Ogien R. La faiblesse de la volonté. P. U. F.; 1993.

Perry JL, Carroll ME. The role of impulsive behavior in drug abuse. Psychopharmacology. 2008;200:1–26.

Pharo P. Raison et civilisation, essai sur les chances de rationalisation morale de la société, Cerf; 2006.

Proust J. La Nature de la volonté. Folio Essais, Gallimard; 2005.

Rachlin H, Raineri A, Cross D. Subjective probability and delay. J Exp Anal Behav. 1991;55:233–44.

Reach G. A novel conceptual framework for understanding adherence to long-term therapies. Patient Prefer Adherence. 2008;2:7–20.

Reach G. Personal view: does an impatience genotype leads to nonadherence to long-term thera-pies. Diabetologia. 2010;53:1562–7.

Reach G. Obedience and motivation as mechanisms for adherence to medication. A study in obese type 2 diabetic patients. Patient Prefer Adherence. 2011;5:523–31.

Reach G, Michault A, Bihan H, Paulino C, Cohen R, Le Clésiau H. Patients' impatience is an independent determinant of poor diabetes control. Diabetes Metab. 2011;37:497–504.

Safford MM, Russell L, Suh DC, Roman S, Pogach L. How much time do patients with diabetes spend on self-care? J Am Board Fam Pract. 2005;18:262–70.

Steele CM, Josephs RA. Alcohol myopia. Its prized and dangerous effects. Am Psychol. 1990;45:921–33.

Story GW, Vlaev I, Seymour B, Darzi A, Dolan RJ. Does temporal discounting explain unhealthy behavior? A systematic review and reinforcement learning perspective. Front Behav Neurosci. 2014;8:76. doi:10.3389/fnbeh.2014.00076.

Wallace M. Future time perspective in schizophrenia. J Abnorm Psychol. 1956;52:240–5.

Wamala S, Merlo J, Bostrom G, Hogstedt C, Agren G. Socioeconomic disadvantage and primary non-adherence with medication in Sweden. Int J Qual Health Care. 2007;19:134–40.

Wolff MC, Leander JD. Selective serotonin reuptake inhibitors decrease impulsive behavior as measured by an adjusting delay procedure in the pigeon. Neuropsychopharmacology. 2002;27:421–9.

Chapter 8
An Intentionalist Account of Doctor-Patient Relationship and Biomedical Ethics

Abstract This analysis may also be relevant for understanding the interaction established between the physician and the patient during their encounter: They must be able to express the contents of their minds in a way that will be intelligible. Indeed when the doctor introduces a novel concept, she extracts it from her own mental puzzle where all concepts are linked in a logical way. This concept may not fit to the mental puzzle of the patient. In the second part of this chapter, our intentionalist model is used to discuss the issue of patient autonomy: An autonomous person is an individual capable of deliberation about personal goals and of acting under the direction of such deliberation. However, this deliberation may be jeopardized by the limits of our rationality delineated in this book. It may therefore happen, at least some time or temporarily, that the patient asks the physician to impose constraints on her. Because the patient comes to see a physician who is also a person with her own mental states, among them knowledge, competence, beliefs, desires, and the emotions that make her capable of empathy. What the patient expects is for the physician to be capable of making the best decisions for her. And when the patient sometimes goes so far as to ask the physician to force her to take care of herself, perhaps she does it because she hopes that this will be done with the cold objectivity of which she knows herself to be incapable.

If the therapeutic relationship consists, for the physician, in trying to understand the patient's *reasons* in order to modify them in a way that she considers good for the patient, can she have access to what is really going on in the head of the person in front of her? Paul Valery imagines Socrates in dialogue with a physician, and Socrates is amazed that the physician knows his body better than he does himself:

> It is strange that you know a thousand times more about me than I do, and that it is as if I were transparent in the light of your knowledge, while I am completely obscure and opaque to myself. (Valéry 1955)

Actually there is a fundamental double asymmetry in the physician-patient relationship. The physician knows my body better than I do, my mind less well then I do (and it is a good thing that we don't have access to what others think. Marcel Aymé showed what would happen otherwise in his play *Les Quatre Vérités*). To the physician, it is my mind that is "completely obscure and opaque", even more so than to myself.

© Springer International Publishing Switzerland 2015
G. Reach, *The Mental Mechanisms of Patient Adherence to Long-Term Therapies*,
Philosophy and Medicine 118, DOI 10.1007/978-3-319-12265-6_8

8.1 Philosophical Analysis of the Doctor-Patient Relationship

So the contents of a person's mental states are the mental baggage whose holistic character we have noted earlier, creating the backdrop for a play with the same title as our book: *Mind and Care*. The key question is then whether it is possible to untangle this complicated web. Because, rather mysteriously, when I mention 'my' mind, I am more certain of what goes on in my head than of what goes on in yours, as if I had a special authority over what is my private life. One of the goals of the philosophy of mind is to try to understand this fundamental asymmetry due to the 'authority of the first person' that we have over our thoughts (Davidson 2001).

It then becomes obvious that the physician, in order to understand why the patient follows (or doesn't follow) his advice, must come to know more of the patient's desires, hopes, regrets, fears, etc. Thus the first questions of a good medical interview must be your first and last name, your age, profession, marital status, whether you have children, how you live, etc. That is why it is so important for the patient to not be perceived as an anonymous individual: During the medical appointment, and why it is important to talk about things other than the disease. Two words are important: *What else?* (Barrier et al. 2003).

This is what patient education experts call *educational diagnosis*. This can be achieved through *concept mapping*, a hierarchical and organized graphic representation of all of an individual's knowledge, starting from a particular concept (Marchand et al. 2002). However, the holism, implying the existence of infinitely many beliefs represents a limit of educational diagnosis: We will never be able to know 'everything' that is in someone's mind, even if we only consider her conscious mental states.

Here, we try to evaluate how a patient will follow her treatment. More precisely, the task is to know whether it is plausible that the patient, with her particular psychology, believes in the effectiveness of a certain therapy and wants to use it whatever its real value. This has nothing to do with the belief that the physician and the patient have, *at the bottom of their hearts*, in its effectiveness itself, a belief that will emerge for both of them out of the estimations of its probability, its credibility and its plausibility. This difference is precisely due to the 'first person authority' that we have over our own beliefs and desires, as Vincent Descombes demonstrates:

> When I want to know whether somebody believes a certain story, I do not in the main ask myself whether the story itself is believable, whether there are reasons to believe it. I ask myself whether it is believable for the person in question, given what I know about his ideas and dispositions. In other words, I ask myself if it is plausible that this person, with his 'psychology', should find this story credible enough to believe it. Things are no different for the attribution of desires. Does the child want this cake? I do not ask myself whether the cake is good or whether there are reasons to find it appetizing, but only whether there are reasons to consider that the child finds it appetizing. By contrast, if I am considering myself rather than someone else, the asymmetry between the third and the

first person that is characteristic of psychological verbs means that now my own reasons are what is at issue. In order to know whether I believe the story you've told me, I ask myself whether the story is plausible, not whether it is plausible that I be in the state of believing it. I look to see whether there are reasons to believe the story, not whether there are reasons to think that I believe it. The same is true for desire: I need to know whether the thing is desirable, not whether the state in which I find myself is one of desiring it. (Descombes 2001, 198–199)

Such an educational diagnosis is possible for two reasons. First, an agent is disposed to express the contents of her propositional attitudes, in assertions or in actions. Thus, I can have access to what you are thinking: I can first listen to what you tell me or observe what you do. But second, there is a much more fundamental reason, pertaining to a characteristic which perhaps defines what is properly human.

8.2 The Principle of Charity

As Davidson says,

we start out assuming that others have, in the basic and largest matters, beliefs and values similar to ours (Davidson 2004, 183),

One can imagine the importance of this aspect of beliefs for the interaction that may and should be established between the physician and the patient during their encounter: To be able to express the contents of their minds in a way that will be intelligible to others, in light of what is called the 'principle of charity'.

According to Isabelle Delpla, it is

a methodological charity, having to do with criteria and principles of understanding rather than the love of one's neighbor. Does that mean that the principle of charity is simply a homonym of Judeo-Christian charity? The answer is tricky, because just as the latter commands us to love others as we do ourselves, methodological charity tells us to understand others using our criteria of truth and reason. If not charity as virtue, this principle is related to the art of credit, to a reflection on what one must accord or not accord to others, at the risk of giving too much or too little, between the pitfalls of prodigality and stinginess. Accordingly, it posits itself as an interpretation norm, the counterpart of laws of hermeneutics in analytical philosophy, taking also translation as a model of understanding. But while hermeneutics' central paradigm is the translation of biblical and ancient texts within historical translation, the principal of charity applies first to a situation of 'radical translation' of a language and culture unknown to us. Otherness is the contemporary one of individuals encountered hic et nunc, rather that the historical one of disappeared authors and texts. It's background is ethnology rather than history (Delpla 2001, 9).

Pascal Engel describes the principle of charity as follows:

The interpreter [the physician] does not ascribe to the audience [the patient] less true and rational beliefs than he ascribes to himself. Using both this data and the ties of holistic dependence between phrases and their mental contents and applying these principles of rationality, the interpreter arrives at a theory of semantic content and of *psychological content*. (Engel 1994, 19)

8.2.1 Four Difficulties

Nonetheless, there are several difficulties that I shall address. First, a patient may be unaware of the contents that some of her propositional attitudes take, even though these still play a causal role in her therapeutic actions. As a result, it would be sometimes necessary to use psychoanalytical techniques to discover these hidden reasons. We saw that the partitioning of the mind proposed by Davidson to explain the phenomenon of *akrasia* does not *necessarily* overlap with the Freudian conscious-unconscious, *but that it can*. Davidson is very clear on this point:

> We can reconcile observation and theory by stipulating the existence of unconscious events and states that, aside from awareness, are like conscious beliefs, desires and emotions. (Davidson 2004, 186)

But even when this is not the case, a second difficulty emerges, due to the fact that the patient may not make explicit the content of her propositional attitudes. As Vincent Descombes notes,

> when I speak, I immediately divide the entirety of what is speakable (of what can be spoken) into two, what I *said* and what I did not say, and what thus became the *unspoken* part of my discourse. Does this unspoken form a totality? No, because it is impossible to combine the total of what is not found anywhere. This scattered category must in turn be divided into two defined groups: the things I didn't talk about because I had *nothing to say* and the things I didn't *want to talk about*. (Descombes 1977, 17)

But unexpressed propositional attitudes may still influence actions. Moreover, what we say and what we really think are often two different things, just as we don't always do what we say that we do. As Pascal Engel reminds us, assertion is neither a necessary condition for the belief (one may believe something without ever expressing it), nor a sufficient condition (one may express a proposition without really believing it). But if an agent acts in accordance with a particular belief, without expressing it, that means the agent agrees with it; in other words, it means that she accepts the contents of this belief. Agreement too is neither a necessary condition for a belief (one may believe something without ever putting it into practice), nor a sufficient condition (one may do things without having corresponding beliefs). Therefore, the best way to come to know the content of propositional attitudes that we think are the cause of a therapeutic action is to see how people act (Engel 1995, 46). To come back to the question at hand, to know if a patient believes that it is good to take her pills, the best way is not to ask her whether she believes it or does it, but *to see* whether she *really* does it. In the first part of the book we noted the difficulty of evaluating patient adherence and the value of electronic pill boxes for this evaluation.

A third difficulty now comes to mind. What happens when the two speakers, the physician and the patient, come from different cultural backgrounds? Is it then still possible to appeal to the principle of charity? Yes: The principle of charity is first and foremost a methodological principle and necessarily transcends the level of background differences; furthermore, we may recall one of the '*infallible principles*' of ethno-psychoanalysis which make possible the practice of a 'transcultural therapy' as proposed by the ethnopsychanalyst Marie-Rose Moro:

> There is an assumption without which ethnopsychoanalysis could not be constructed, that of psychic universality, i.e., of a fundamental unity of human psyche. From this assumption follows the requirement of giving the same ethnic and scientific status to all human beings, to their cultural and psychic production, to their ways of living and thinking, however different and sometimes confusing they may be [...]. Another universal human characteristic is of course the fact that everyone has a culture and this is perhaps the foundation of one's humanity and universality. (Moro 2002, 164)

Of course this does not mean that a patient who has a different cultural background than me believes the same things as I do. But she believes in the same way as I do, and her beliefs influence her actions through a mechanism like the one that is at work in my mind.

Finally, a fourth difficulty arises when the physician is treating a patient who does not speak her language. It depends on the interpreter in this case, and the exchange between the holisms of the physician and the patient, which is already an interpretation, runs the risk of being confused by another level of interpretation.

As we saw earlier, the adherent patient (a) has at least *some* attractive representations of the future of her treatment, (b) is able to access these representations when she is weighing her adherent actions against nonadherent ones, and (c) abides by the principle of foresight, so that these attractive representations prevail and adherent actions are thereby chosen. The therapeutic alliance aspires to develop attractive representations of adherence and inspire in the patient a commitment to the principle of foresight. Let this be our outline for therapeutic alliance.

8.2.2 Back to Patient Education

Here we have reached the crucial point that will reveal the efficacy of the doctor's prescription. We can effectively deduce from it the conditions under which one of the doctor's beliefs may be transmitted to the patient, in a way that will lead her not only to accept it, but also so that it will acquire the essential quality of durability, the ability to produce adequate actions over the long run. It is thus essential to think about the conditions that will make this sharing of beliefs efficient. And this is tantamount to thinking about conditions that will make patient education efficient (Reach 2009a).

Several factors may affect the sharing of beliefs. The first factor is the insistence put into the transmission of the belief—we cannot hope that someone will persistently believe something if the object of the belief is presented to her in passing, or only alluded to: Patient education takes time. Most medical visits involve the introduction of a novel concept to the patient, and it is important that the healthcare provider explains the concept and checks that this has been understood, and, if necessary, clarifies and tailors the information to the patient (Schillinger et al. 2003). There are three barriers to an efficient communication (Reach 2009b):

(1) The patient may not understand *the words*, either because she speaks another language, or because the doctor uses a jargon; (2) She may not understand *the message*, because medical concepts are often only understandable by doctors; (3) The concept may not fit with her psychology and her culture. Indeed when the doctor introduces a novel concept, she extracts it from her own mental puzzle (see Chap. 3), where all concepts are linked in a logical way. This concept may not fit to the mental puzzle of the patient (Fig. 8.1).

The second factor involves 'external' conditions: The patient's attention. One isn't likely to succeed in sharing a message if there is disruptive ambient noise, like loud drilling from the next room. One must create proper conditions for education. For children with diabetes, summer camps are perhaps a particularly suitable environment.

Moreover, a physician must know how to seize the golden moments for education. One obvious such opportunity is when the patient asks for an explanation pertaining to her treatment; for this means she perceives a 'gap' in the holistic puzzle of her beliefs, and she would like it filled. The third factor is trust. In order to take on a new belief at another's exhortation, I must trust the exhorter. Trust must be personalized: The physician can't simply *hope* that a patient will believe her— they hardly know one another. This is perhaps one reason why messages are better transmitted in doctor-patient relationships of long standing. The fourth factor is the effect of the degree of probability, plausibility, and desirability of what you are invited to believe. The fifth factor is the utility of the content of this belief, because one believes that which corresponds to a desire, and one desires above all what is

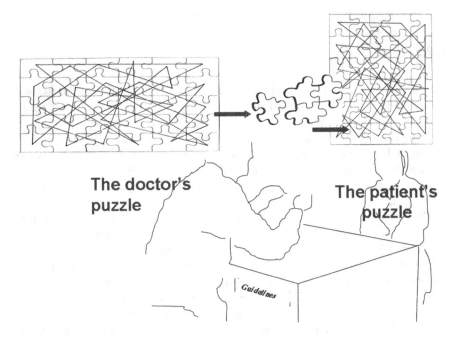

Fig. 8.1 Patient education, a communication between two mental puzzles

good, and what is good is above all useful—"By good I will understand what we certainly know to be useful to us." Spinoza, Ethics, IV, Definition 1—This is how the new belief will integrate into the mental puzzle.

8.2.3 Empathy

It is here that the physician must show herself capable of *empathy*. According to Carl Rogers' definition, this means

> to perceive the internal frame of reference of another with accuracy and with the emotional components and meanings which pertain thereto as if one were the person, but without ever losing the "as if" condition. Thus, it means to sense the hurt or the pleasure of another as he senses it and to perceive the causes thereof as he perceives them, but without ever losing the recognition that it is as if I were hurt or pleased and so forth. (Rogers 1959)

The idea of empathy is at the root of person-centered therapy, developed by C. Rogers (see Prochaska 1994, 133–162) and of the general concept of counseling, which is used particularly to improve adherence in illnesses such as AIDS.

And when the patient perceives in the physician the desire to show empathy towards her, perhaps she will feel the symmetrical desire to attempt to share beliefs. This description of the doctor-patient encounter finds an echo in the description of the therapeutic relationship by Michael Balint:

> It is on this basis of mutual satisfaction and frustration that a unique relationship is established between a general practitioner and those who stay with him. It is very difficult to describe this relationship in psychological terms. It is not love or mutual respect, or mutual identification or friendship, although all these elements are present in the relationship. We have called, for lack of a better word, a 'company of mutual investment'. What we mean by this is that the general practitioner progressively acquires precious capital invested in his patient and, respectively, the patient acquires a precious capital which he deposits in his doctor (Balint 1988, 265).

The sharing of beliefs implies a dialogue aimed first at discovering the other's beliefs concerning her representation of illness and health, her expectations and fears concerning the treatment and the physician. This conversation is the beginning of a negotiation, which will lead to a necessary therapeutic alliance between the physician and the patient.

8.2.4 Therapeutic Alliance

Here we can quote an enlightening passage from Jean-François Bloch-Lainé's text on therapeutic alliance:

> The conditions required to achieve a therapeutic alliance when treating drug addiction are, at first glance, the same as the conditions generally required in any medical situation: the patient is identified as ill, the therapist is identified as the one offering treatment.

> The therapeutic alliance between the patient and the physician can exist only as long as there is no contempt, no complicity, no indifference or complacency, no hate or love, but only genuine mutual respect. The respect felt by the patient towards the physician cannot be dissociated from the required trust that the patient has in the physician, everyone agrees with this banality. The respect of the patient felt by the physician is more difficult to formulate. To respect the patient supposes that one respects the person treated and the reality of her illness (Bloch-Lainé, Internet).

The first part of this text refers to empathy, according to the definition given earlier. One wonders if it is not easier to formulate the physician's respect for her patient if the former recognizes the legitimacy of the patient's perspective on the illness in addition to the medical perspective. Is it not a display of empathy to show the patient that her perspective on the illness is legitimate? This may lead the patient to accept that the physician's perspective is legitimate as well. Trust, in the context of the exchange of representations, becomes the basis for mutual respect.

8.2.5 Patient's Beliefs, Physician's Beliefs

Here it is useful to clarify more precisely the concept of belief. Following Pascal Engel, there are several types of beliefs, defined according to the degree of objective guarantee attached to the representation and according to the degree of subjective confidence the agent has as to the truth of this representation. Thus we can speak of *hypotheses* or *conjectures* when the beliefs are likely to be true or to have an objective basis, or when they are being verified; of *convictions*, when there is a strong subjective feeling but its basis is not guaranteed; of *prejudices* and *superstitions* when the objective guarantee of the opinion is very weak or nonexistent, although the agent may feel a very strong conviction to the contrary; and finally of *confidence* (in someone) or *faith* (in something) when, despite a very weak objective guarantee, the degree of subjective certitude is very strong and goes further than what the facts and guarantees allow one to affirm (Engel 1995, 10–11).

Using this classification, we can distinguish those that concern more or less verifiable facts and those that concern a priori unverifiable representations, because they concern the future, which is essentially contingent.

First, there are matters of fact of which patients may nonetheless feel uncertain. Patients' knowledge often includes some uncertainty and must be seen as beliefs of one of the two first categories, hypotheses, conjectures or convictions, as described above. This does not mean that their propositional content, i.e., what they believe, is not true. But we can also suppose that the degree of certitude plays a role in the belief's motivational force. We have studied the effect of the degree of certitude concerning diabetic patients' knowledge on the adjustment of insulin doses (Reach et al. 2005).

Recall that when a belief is part of an action's reason/cause, it is typically a belief about the outcome of that action. For instance, the active belief in the intention to drink tea in order to alleviate urinary burning is a belief about the action effected: Namely, it's the belief that the action will have the desired effect. Since these beliefs concern the future, and since the future is uncertain, such beliefs are also of the second type. They belong more to the last two categories of beliefs, trust or faith, but also prejudice or superstition.

But we can note that the same may be true of numerous medical ideas, which, not having been proven and the physician not being able to affirm that they are certain, are deep down merely beliefs on which they base their therapeutic strategies. This is what justifies the emergence of evidence-based medicine, which aims to eliminate uncertainty from the development of therapeutic strategies, or, more precisely, to quantify the uncertainty. It uses large scale controlled clinical studies and meta-analysis of all the available information produced by these studies: The principle of statistical demonstration by a controlled study is to limit the risk of error by depending primarily on probability claims. Ever since the *DCCT* study, I don't simply *believe* that control of diabetes lowers the risk of complications, I think that I *know* it; thus I feel more authorized to propose this therapeutic strategy to you. The advent of evidence-based medicine could come down to applying Ramsey's principle, stipulating that the degree of truth of a belief represents the probability that it will lead to a successful action.

The preceding nonetheless implies that the sometimes antagonistic relationship between physician and patient is not a conflict between knowledge on one side and beliefs on the other, but rather a confrontation between two webs of beliefs, i.e., between two holisms. The patient's web of beliefs is greatly shaped by her past and her personal experiences. That of the physician is vaster and better documented, but only in what concerns the prediction for the *future* of an *actual* case. The physician *believes* that she can make an analogy between what she *knows* about what *happened* in the past with other patients, whether that experience be personal or from books, and from the evidence-based medicine, although this evidence is essentially statistical.

According to Hippocrates:

> Physicians come to a case in full health of body and mind. They compare the present symptoms of the patient with similar cases they have seen in the past, so that they can say how cures were affected then. But consider the view of the patients. They do not know what they are suffering from, nor why they are suffering from it, nor what will succeed their present symptoms. Nor have they experience of the course of similar cases. Their present pains are increased by fears for the future. They are full of disease and starved of nourishment; they prefer an immediate alleviation of pain to a remedy that will return them to health. Although they have no wish to die, they have not the courage to be patient. Such is their condition when they receive the physician's orders. Which then is more likely? That they will carry out the doctor's orders or do something else? (Hippocrates, Prognosis)

The physician must take into account her patient's past, present, and future. The Persona® test which places people in the categories of "analyzer, facilitator, promoter, controller" suggests that the 'facilitating' individuals are those who face at the same time the past, the present and the future. It is in this category that we find most doctors...

It is because the physician also has this vision of the future that she can act on her patient's behalf *according to her own principle of foresight*, and we will precisely see in the next chapter that failing to abide by this principle may represent a cause of doctor's clinical inertia. And her most difficult task is to help her patient acquire, little by little, the principle that will lead her on the road to adherence. In so doing, the physician fosters acceptance of propositional attitudes that form the holistic richness of her mind, the facts that will protect her future rather than those, often more seductive, that would allow her to enjoy the present moment.

It is precisely here that there is a difficulty due to a difference between the functioning of the physician's mind and the patient's: If we return to the concept of time discounting described earlier, we can calculate that the discount rate and its curve *must sometimes be different* for the physician and the patient.

Steven Feldman clearly showed the reasons for these differences (Feldman et al. 2002). For instance, because the physician has chosen the medical profession and practices it, she likely has a weaker discount rate for the future than the patient, or attributes more importance to health. In short, physicians may be more able to delay gratification and overplay the importance of health. In particular, patients from cultures where the present moment is important might have a very different future discount rate. Certainly, we have insisted on the fact that the operational mode of thinking is universal: The role of beliefs, desires and emotions in the genesis of actions is certainly the same for all humans. But this is fundamentally true only from the qualitative point of view, and quantitative differences—say, for the future discount rate—may have important consequences.

We can mention two conditions where the discount rate of the patient and the physician are different: Patient's social deprivation and serious depression. Here, the feeling of the future has more or less disappeared (the future discount rate approaches infinity) and a physician who proposes a treatment of indefinite duration is sure to fail. This obviously does not mean that all therapeutic attempts are impossible; it merely illustrates the importance of taking into account considerations of this type. In this article, the authors suggest that the discount rate might not play an important role in the short term. This may explain the higher frequency of adherence in cases of acute diseases.

8.2.6 *The Therapeutic Relationship*

Two aspects of the therapeutic relationship emerge at the conclusion of this analysis. First, we saw that one cannot *decide*, all of sudden, to believe something, just as 'one decides to go away for the weekend' as Pascal Engel puts it; on the other hand, one may believe what one is told—the patient believes what the physician tells her: As we saw, the sharing of beliefs (and desires) between the physician and patient is possible. Caring is sharing: Saint Martin doesn't give his coat to the beggar, he shares it.

This sharing does not concern only beliefs and desires. It can also include emotions. Here we can again quote Pierre Livet:

> the sharing of emotions plays a double role here. On the one hand, it is an important emotional consequence of the preference being tested. To discover that others do not share the emotions connected to our preference provokes a strong negative emotion. Conversely, the fact that others place a certain value on it, which we have not considered, is an important

source of emotion and revision of our preferences. On the other hand, it is an empirical ersatz of universalisation. If we can suppose in advance that our emotion is not shareable, we cannot give value to our preference, even if for us it is a deeply rooted preference (Livet 2002, 185).

What this remark suggests is that the patient and the physician, during their interaction, are led to confront arguments of an emotional nature. For example, when the patient invokes her pleasure of smoking, the physician fears the appearance of long-term complications if the patient does not give it up. And according to the definition of empathy, one must be capable of feeling what the other one feels. In the next chapter, we will revisit the very concept of empathy and show how, paradoxically, its use can lead to clinical inertia. To try to understand someone else's emotions means checking whether she is ready to revise her preferences. For Livet, the debate about preferences is actually a debate about values. If the patient is resistant to the idea of a revision of preferences, that means that she places more value in smoking than in the idea of health. The first role of the physician is then to demonstrate to her the value of health, for which the physician is a sort of representative. That is why the physician certainly has the right to smoke, but smoking in front of patients is highly reprehensible. Health might be seen as a universal value, or even the archetype of what is valued: According to Canguilhem,

for man health is a feeling of assurance in life to which no limits is fixed. *Valere*, from which value derives, means to be in good health in Latin. Health is a way of tackling existence as one feels that one is not only possessor or bearer but also; if necessary; creator of value; establisher of vital norms (Canguilhem 1989, 201).

But often the discussion falls short and the patient does not admit the necessity of revising her preferences. Then we come to a conflict, and nonadherence is likely to be its expression. I will show that in fact it is not a conflict between the physician and the patient. It is a deeper issue of a conflict between two current trends in medicine: One that consists in desiring to benefit from the progress of modern science and 'to improve adherence', in the name of the patient's well-being, and one that wants to promote patient autonomy.

8.3 Adherence and Autonomy

Up to now we have assumed that the physician's desire to improve adherence is self evident. In a recent report by the World Health Organization (2003) devoted to adherence in cases of chronic disease we read that finding a solution to nonadherence would be more beneficial than any other medical advance. However, at a time when medicine is becoming more and more effective and sometimes even manages to prove its effectiveness according to the criteria of evidence-based medicine, society is becoming more sensitive to patient rights and autonomy. There looms a possibility of conflict created by medicine itself: What attitude to adopt when the physician's and the patient's points of view diverge? How far can the physician go in her desire to convince the patient?

Etymologically, a person is autonomous if she chooses herself (*autos*) which rules (*nomos*) she is going to follow and applies them, in the same way that an autonomous government writes its own laws and has the freedom to enforce them. This supposes that the person, like the government, intends to have control over her actions and does not give the right to control her action to anyone else without express permission. Analyzing the notion of autonomy then comes down to analyzing the control that the person has over her actions. And inasmuch as an action can be characterized by its intentional nature, analyzing the notion of autonomy comes down to analyzing the control a person has over her reasons and her capacity to act according to these reasons. More precisely, the autonomous person has the capacity to perform an action or to not perform it, or to perform one action rather than another. Thus she is endowed with the capacity to choose. However, we must still clarify whether we are speaking of an autonomous person or an autonomous action. Indeed, there are persons whom we would not consider autonomous who nonetheless perform autonomous actions and, inversely, autonomous persons who from time to time perform non-autonomous actions.

When we go from the generic concept of personal autonomy to the specific concept of personal autonomy in the context of a medical decision—making a choice affecting one's health—the concept is somewhat modified. Call the modified concept *therapeutic autonomy*. We are now no longer considering only the person. Into the discussion is introduced the fact that the person is engaged in a therapeutic, or binary, situation. In this situation, the person *became ill* and comes to consult a doctor; she has become a patient who will be dealing with a medical team: The question of autonomy now needs to be considered not only from the point of view of a person, but of a person in a therapeutic relationship with another. The person will have to be autonomous not only in relation to herself, i.e. to have or not have control over her actions, but also in relation to the physician, to another, who may want to control her actions (in fact, the relationship may be more complicated than simply binary between the patient and the physician, as it may involve, for instance, the patient's family and friends, who might also want to intervene in the decision making process).

This has a major implication: Recognizing that the patient, as a person, is an autonomous being implies for the physician a particular behavior in regard to her autonomy. And, in an attempt to benefit the patient, this behavior can override the patient's decision (this is often called 'paternalism'); or, it can seek to respect the patient, which for now can be termed 'autonomist behavior of the physician'.

8.3.1 Therapeutic Autonomy in Medical Ethics: Fourth or First Principle?

The respect of patient autonomy has become a major principle of contemporary medical ethics. It has been proposed that we should add to the two Hippocratic principles of non-maleficence (*primum non nocere*) and beneficence (act for the welfare of the patient) a principle of justice (to guarantee an equitable repartition of medical resources) and a principle commanding respect for patient autonomy (Beauchamp and Childress 2001).

The principle of autonomy has important consequences for the contemporary practice of medicine. It is specifically at the origin of the notion of patients' informed consent and, more generally, of patients' rights, now recognized by law. This is the case for biomedical research (in France, regulated by the law "Huriet"), and for therapy in general. Thus, the French law of March 4th 2002, article L.1111-4 declares that:

> Each person makes decisions concerning her health, together with the healthcare professional and in view of the information and prescriptions provided. The physician must respect the patient's decision after informing her of the consequences of her choices. If the person's decision to refuse or stop treatment puts her life in danger, the physician must do all that is possible to convince her to accept the necessary treatment. No medical act or treatment can be performed without the free and informed consent of the person and this consent can be withdrawn at any time.

In other words, the law grants the patient the right to be nonadherent.

So the notion of therapeutic autonomy presupposes that what is at stake is the patient's capacity to freely make choices concerning her health, and it is this capacity that must be respected according to the fourth principle of medical ethics. But just as it is not self evident that trying to improve adherence is legitimate, it is also unclear whether therapeutic autonomy is even *possible*. We must then analyze what is included in the concept of therapeutic autonomy. In the law we just quoted, for example, we must discern the appropriate meanings of the terms 'person', 'decision', 'will', 'choice', 'free', 'informed', etc. It then becomes not so much a question of ethics or law, but of psychology, or rather, as soon as we ask whether therapeutic autonomy is *possible, it is again a philosophical question.*

8.4 Philosophical Conception of Autonomy as a Reflective Activity of the Mind

In a psychological, or philosophical, conception of autonomy, the autonomous individual is capable of choosing according to her values and preferences which are an expression of her individuality: For instance, a patient could choose to quit smoking because she places more value in the idea of health than in the pleasure of smoking. Hence it is the notions of value and preference that it is now time to analyze: What does it mean to assign a value to something, which will give it a certain weight during decision making?

8.4.1 Reflective Activity of the Mind

Figure 8.2 illustrating our intentionalist model of adherence, so-called second order mental states play a vital role: Beliefs about beliefs, desires about desires.

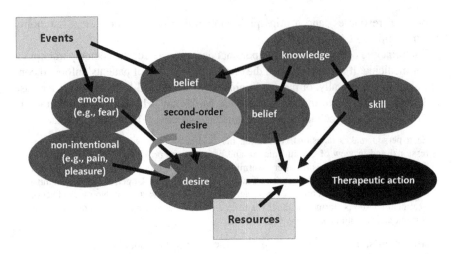

Fig. 8.2 Desire about desire, the reflective activity of the mind

There is then a possibility of *a reflective activity of the mind*, which, as we shall see, is the condition of autonomy. This is precisely stated in the Belmont Report's definition of autonomy:

> An autonomous person is an individual capable of deliberation about personal goals and of acting under the direction of such deliberation.

This notion of second-order desire dates back to the philosophical works of the 70's and 80's, the years that saw the birth of contemporary bioethics. These works agreed on the importance of an individual's reflective activity concerning her own desires and on the use of the concept of value for a psychological definition of autonomy. Thus Lewis (1989), in his article 'Dispositional Theories of Values' suggests that the value that one attributes to something and which will make us give a preference to it can be understood as *wanting to want* this thing: A smoker may want to smoke a cigarette and at the same time not ascribe any value to her addiction and actually want to quit smoking. It is only if she *wants to want* to smoke that we can say that she places a value on smoking. Gerald Dworkin, in *The Theory and Practice of Autonomy*, published in 1988, uses this reflective activity to define the autonomous person. Autonomy is a second-order capacity to reflect critically upon one's first-order preferences and desires (Dworkin 1988). In addition, autonomy is the ability to either accept one's preferences desires and wishes or try to change them in light of higher-order preferences and values.

Not only does this capacity define the autonomous person, it also defines the very concept of person; so argued Harry Frankfurt in 1971. To be a *person*, and not simply the puppet of brute desires (what he calls a *wanton*), is to have the capacity for reflective self-evaluation manifested in the formation of second order desires which supercede first order desires. What's more, I am a person if it is important to me that my will be free, i.e., that I be capable not only of wanting something and doing what I want, but being able to do what I want to want (Frankfurt 1988, 11–25).

The notion of mental states of different orders obviously presents the risk of infinite regression. If I act according to my second-order desire and not according to my first-order desire, why couldn't there be a third-order desire that I should also consider, etc.? Frankfurt (1988, 159–176) considers this question in his essay *Identification and wholeheartedness*. In fact, my 'decision' to act according to my second-order desire means that I have no need for higher-order desires, just as when I make a calculation, I could infinitely keep checking whether it is correct. If, at a certain point, I decide to stop checking, that means I decide that no additional verification is necessary.

Frankfurt likens the notion of person to that of free will founded on the reflective activity of the mind. He does not use the term autonomy in his article, written before ethicists became interested in this subject. But later, Dworkin equated the reflective activity of the mind with the notion of the autonomous person: This could suggest, transitively, that to respect autonomy boils down to respecting the person, something we will reexamine at the end of this chapter.

8.4.2 An Intentionalist Analysis of Autonomy

Thus, an autonomous person is one who chooses an action according to her values and preferences. To ask whether therapeutic autonomy is possible is to ask whether a person can control the interferences of various mental states. We will ask this question using the theoretical framework of the intentionalist model of adherence proposed in Fig. 8.1.

According to this model, it is clear that skills have only an instrumental role in the performance of a therapeutic act. The same goes for knowledge, used mainly to form beliefs which will confer on desire its pro-attitude role during the performance of the action: I believe that exercising is one of the actions which will make me lose weight, and because I have this belief and this desire, I go to the pool. And I believe that exercise is one of the actions that makes one lose weight because I know that one loses weight when burning more calories than one ingests. Patient education for the treatment of chronic diseases, information regarding therapeutic decision-making, can give me the knowledge, the beliefs and the skills necessary to enact my beliefs. But the indirect, backstage role of knowledge and skills are insufficient, by themselves, to produce autonomous therapeutic action. We must consider what is front and center: Beliefs, and most especially, emotions and desires.

We saw earlier that the formation of a new and enduring belief (for example, 'I believe that I am ill') takes shape after a series of evaluations: That of probability, taking into account everything I think besides; that of plausibility, searching whether I can explain the phenomenon that is the object of my belief; and finally, that of credibility, founded on the evaluation of the reliability of the sources available to me. However, this evaluation can be biased by our frequent use of heuristics: Heuristics are simple, fast and efficient procedures that help find adequate, though often imperfect, answers to difficult questions and they explain how we can make decisions in a context of uncertainty. Indeed, our way of thinking is more often "fast" than "slow" (Kahneman 2011). Furthermore, the Prospect Theory predicts that the evaluation of outcomes can be biased by the loss aversion effect (Kahneman and Tversky 1979).

We have also noted that cognitive phenomena are passive and that we cannot simply believe at will; this does not mean, however, that we cannot believe what someone else tells us. Nonetheless, we see here a difficulty for autonomy, as the formation of a belief can depend on another person, i.e., on the degree of dependence towards another person. And manipulation is possible specifically in the domain of therapeutic autonomy, while one is collecting information leading to 'informed consent'.

We have also insisted, following Pierre Livet, on the role of emotions in the revision of our beliefs, desires and preferences. This could suggest that in as much as emotions are a response to events and have an influence on our beliefs and desires, we cannot have complete control over our desires and beliefs. And as concerns the strength of our desires, we saw that it depends on the proximity of the reward to which the desire refers: The impatient ones, who have a high rate of future discounting, run the risk, through the phenomenon of preference reversal, of placing more value on immediate rewards than on health. Although we are incapable of directly controlling our degree of impatience, it can be modified indirectly by factors (some of which we can control, some not): We saw how the discount rate rises for example under the influence of social deprivation or depression, or simply after drinking alcohol.

Finally, one last element that could make the exercise of autonomy difficult is the influence of unconscious reasons. We know that our mental states can be conscious or unconscious, hence the image of the mind as composed of multiple selves. In this case, if autonomy is defined as the mind making choices according to its own rules, which of the 'selves' will override the others to impose its rule?

So, after all, is therapeutic autonomy *possible*? Is there anyone who can declare, like Augustus, that *I am master of myself as I am of the Universe*? We are tempted to answer that therapeutic autonomy is *possible, but it cannot always be exercised*. And to paraphrase a famous line, we propose that everyone can be autonomous sometimes, and some people, doubtless very few, can be autonomous all the time, but it can't be that everyone is autonomous all the time. I will now support this theoretical claim with empirical data.

8.4.3 *Empirical Data: Patients Do Not Always Wish to Exercise Their Autonomy*

Practicing physicians are all familiar with the patient who refuses to make autonomous medical decisions. Such patients insist "Doctor, do whatever you think is best" or they defer to their spouse or children. Such realities can conveniently disappear in the theoretical literature; yet they fly in the face of the current reverence for patient autonomy.

One study showed that patients often prefer that the physician make the decisions, especially when the illness was severe (Ende et al. 1989); indeed, the same was true even when the patient herself was a physician (Ende et al. 1990). We can recall the case of Ingelfinger (1980), the eminent editor of the *New England Journal of Medicine* who, in a famous article to which we shall return, relates his disarray during a period of uncertainty after learning he had cancer: He didn't know what to do, and then someone told him: What you need is a doctor (who will make the decisions for you)!

There is a wealth of literature in the domain of oncology (Lee 2002), and also regarding other chronic diseases such as heart disease (Mansell et al. 2000) and asthma (Adams et al. 2001), where it has been shown that patients want to participate in making key decisions but not in minor decisions. A study of patients with high blood pressure has shown that physicians underestimate their patients' interest in information and simultaneously *overestimate* their desire to participate in medical decisions (Strull et al. 1984).

Another study found that only 20 % of patients choose to have an active role in medical decisions, the others playing a collaborative (40 %), or passive (40 %) role (Doherty and Doherty 2005). In an American study involving more than 2,500 participants, 96 % of patients asked to be informed and wanted to have several therapeutic options offered to them, 52 % of them preferred for the physician to make the final decision, and 44 % preferred to rely on the physician for the information rather than doing the research themselves (Levinson et al. 2005). In sum, this empirical data confirms that not all patients, and not all the time, wish to exercise their therapeutic autonomy, which is compatible with the idea, defended here, that it *cannot* be always exercised.

I will come back at the end of the chapter to the question of whether this situation is in fact unfortunate. In other words, we shall ask whether we can—and should—train patients to become autonomous. But first, I am going to discuss the implications of this psychological reality for the therapeutic relationship.

8.4.4 Therapeutic Autonomy and Models of the Patient-Physician Relationship

How we understand therapeutic autonomy determines how we view the therapeutic relationship. Nowhere is this more apparent than in the treatment of chronic diseases. The management of chronic disease demands a switch from physician authoritarianism—which is sometimes justified in acute care—to a relationship in which the physician takes into account the patient's autonomy. Several models of the therapeutic relationship have been described accordingly. I will refer here to the classic article by Emanuel and Emanuel (1992) (Fig. 8.3).

The *paternalistic model* is the traditional one. With the goal of assuring the patient's health and well-being, the physician uses her expertise to evaluate the

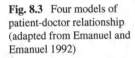

Fig. 8.3 Four models of patient-doctor relationship (adapted from Emanuel and Emanuel 1992)

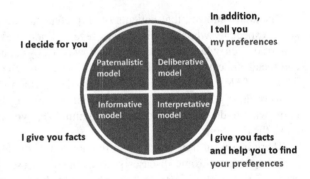

patient's condition and accordingly decides what the most appropriate treatment is. She gives selective information to the patient in order to obtain the patient's consent.

> In the tension between the patient's autonomy and well-being, between choice and health, the paternalistic physician's main emphasis is toward the latter. […] The conception of patient autonomy is patient assent, either at the time or later, to the physician's determinations of what is best.

On the other extreme, there is the *informative model* of the physician-patient relationship: The physician's responsibility is to give the patient all the relevant information concerning the diagnosis and the different therapeutic options without hiding from her the possible uncertainties, so that the patient can choose a therapeutic option. The physician's role is then to help her execute it. Here,

> the conception of patient autonomy is patient control over medical decision making.

Thus, it is believed that the patient has preferences; what she lacks is data. The physician's role is to provide the patient with data, and then the patient will choose according to her preferences. In this model, the physician's preferences, what she knows about the patient's preferences, and her judgments concerning these preferences are discounted.

In the *interpretative model*, the goal of the therapeutic relationship is not only to give the patient information, but to help her elucidate her preferences, as they may not be explicit, and then to help her act according to her preferences by explaining which therapeutic option is the most appropriate is the most appropriate given those preferences as revealed during the therapeutic interaction. Significantly, the physician does not influence the revelation of the preference. Here,

> the conception of patient autonomy is self-understanding.

Finally, in the *deliberative model*, the goal of the physician-patient relationship is to help the patient determine and choose what seems best for her health. Unlike the previous models, the physician indicates which options she finds preferable for the patient's health. In the course of their discussion, the physician and the patient express value judgments concerning the different choices. This discussion is carried exclusively on the medical plane, never on a moral plane; we are dealing with persuasion, not coercion.

> Not only does the physician indicate what the patient could do, but, knowing the patient and wishing what is best, the physician indicates what the patient should do, what decision regarding medical therapy would be admirable. The conception of patient autonomy is moral self-development; the patient is empowered not simply to follow unexamined preferences or examined values, but to consider through dialogue alternative health-related values, their worthiness, and their implications for treatment.

It is easy to criticize the paternalistic model, which appears to have proved its own ineffectiveness: As we saw, when the physician is the only one responsible for the medical decision (as in the paternalistic model), patients follow the advice, at best, half the time; so it must be a misguided way of giving advice. This criticism of the paternalist model has led to the development, particularly in diabetes care (Funnell et al. 1991), of a model close to the interpretative model, known as the *empowerment* model. Bob Anderson and Martha Funnell, its principal promoters, define it as follows:

> The patient empowerment approach to diabetes patient education seeks to maximize the self-management knowledge, skills, self-awareness and sense of personal autonomy of patients to enable them to take charge of their own diabetes self-management. Empowered patients are those that have learned enough about diabetes and themselves, so that, in consultation with health care professionals, they can select and achieve their own goals for diabetes care (Anderson et al. 2000).

This approach is nearly identical to the interpretative model described earlier.

It is interesting to note that for Anderson and Funnell, as long as the therapeutic program is chosen by the patient, the concepts of adherence and nonadherence become 'dysfunctional', inappropriate and useless. Consequently, it is no longer necessary to wish to 'improve adherence' since it refers to the physician's chosen program, while the empowerment approach consists of making sure the patient chooses the program first with autonomy.

We may ask whether this position is realistic. Several reasons suggest that it is not. (1) Contesting the very existence of nonadherence in this way seems to be a sleight of hand that ignores common sense. (2) Is it not naïve to believe that once the patient chooses her treatment, she is going to follow to it? This is simply forgetting that, as we have shown, desires are mental states that can change with events and emotions; and it ignores well-known phenomena such as weakness of the will or preference reversal. (3) This position does not take into account that individuals differ in their *Locus of Control*. That is, for some, one's health is controlled by internal factors, which depend on the individual; for others, external factors, e.g. the physician, spirits, chance, are determinative (Rehault 1982). (4) If the patient's mind is comprised of multiple selves, with a possibly unconscious component, which 'self' is going to be 'empowered'? (5) Last but not least, we saw that some patients cannot be autonomous; and some know that their will is weak. They explicitly say: *I can't help it.*

8.4.5 Freely Giving up Autonomy

We have suggested that accepting a long-term treatment implies that we embrace a principle of foresight enjoining us to give priority to the future. Jon Elster mentions that just as it is impossible to decide to become immoral, so it is highly doubtful that one can decide to have foresight, because wanting to take into account the future means one *already* has foresight. You have it or you don't: You cannot force yourself. Elster notes that even the best magician cannot tie himself up perfectly, because there comes a points when he does not have sufficient strength to tighten the last knot:

> Typically, A needs the assistance of another agent B to bind him. Ulysses had to ask the sailors to tie him to the mast (Elster 2000, 276).

Here we see the second aspect of the therapeutic relationship: It may happen, at least some time or temporarily, that the patient asks the physician to impose constraints on her—to cinch down the final knot.

This conclusion may seem particularly provocative as it appears to contradict a certain tendency of contemporary medicine towards patient autonomy. Nonetheless, the ambivalence in the physician-patient relationship explains why a physician cannot treat herself, gives its meaning to that last consultation the patient asks for before going on vacation, to the word 'prescription', to the search for help from others in Weight Watchers, Alcoholics Anonymous and the like. This is also the case with the use of gastric banding in the surgical treatment of obesity. By reducing the size of the stomach, it imposes a physical constraint that keeps the patient from eating as much as she wants. The gastric ring or stomach binding is explicitly seen by many patients as a way to fight the consequences of the weakness of the will.

If the medical profession is to remain relevant, limits to patient responsibility must be acknowledged. Michael Balint (1957) noted that

> patients should be educated to mature responsibility towards their illness; but it is necessary to add a rider; with certain outlets for dependent childishness. As so often in medical practice, here too the problem is that of proportion; how much maturity should be demanded, and how much childlike dependence on the doctor tolerated? (Balint 1988, 239)

One could say that we have here an old fashioned paternalistic vision of medicine. But is it not based on an inescapable psychological reality? A number of diabetic patients counted on a physician or a nurse. Of course, we have here a state of affairs that needs to be improved by promoting patient education. But it would certainly be a mistake to fail to recognize that at least in certain cases, patients cannot help but rely on their physicians.

The idea of a therapeutic alliance does not imply that the physician should discard her responsibility altogether: Once the physician abrogates responsibility for the health of the patient, patients will seek elsewhere for the help they desire.

Denying the need for prescriptive guidance does not make the need disappear; it only drives patients to seek "alternative medicine" providers—a growing trend in North America and Europe.

In *Une Pédagogie de la guérison est-elle possible* (1978), Canguilhem (2002, 93–94) wrote:

Nothing is more common and more profitable these days than an anti-x proclamation. Anti-psychiatry was the beginning and anti-medicine followed. A long time before Ivan Ilyich's exhortations to the individuals to take back the regulation of their health, self governance of their healing and control of their death, the psychoanalytical and psychosomatic fallout to the level of vulgarization proper to the media has popularized the idea of a desirable transformation of the patient into his own doctor. It was then believed that the eternal theme of one's physician was being taken up again. As times are hard and a niche difficult to find, a growing number of practitioners of non-scientific therapies – science is the enemy – boast of being able to achieve what they reproach the physicians of neglecting and of failing to do. Whence the call to disappointed patients: come tell us that you want to be well, and together with you we will do the rest. (Canguilhem 2002, 93, 94)

We must be able to hear the patient's paradoxical demands and not take the risk of transforming the patient into her own physician, only to become a physician in spite of herself. I am thinking in particular of the following situation, frequently encountered in diabetes care: Patients in danger of serious complications, who are not adherent, who do not have the necessary exams done (annual eye exam, trimestrial measurement of glycated hemoglobin), or who do not perform the daily tasks of the treatment (measuring the blood sugar, keeping a logbook, or even taking all the daily injections of insulin), and *who nonetheless come to the consultations*. Why do they come?

8.4.6 One's Own Physician: Healing One-Self

Here it might be interesting to take up the grammatical analysis performed by Vincent Descombes in his *Complément de sujet*.

The reflexive form of a verb like 'to take care of oneself' allows us to understand how the physician Callias takes care of the patient Callias, how he takes care of himself. The relation of Callias to himself is the relation to Callias as a human being who happens to have medical knowledge (i.e., a power that he can use, presenting himself as an agent) and also who has the required disposition to heal following a treatment (the object of care). Such action is necessarily reflexive, but that does not mean that the agent is in a relationship with anything other than a patient (object).

This means that the physician Callias, if he took care of himself, would not be doing anything other than his job with someone who just happens to be himself.

The care is not offered to the physician as such, but to a man who happens to be a physician. The identity of physician and patient is contingent (Descombes 2004, 110).

And, as we saw, physicians generally do not treat themselves. Why? Because, in reality, it would be most often the patient Callias (and not the physician Callias) who would be treating himself. But then he would not be performing a medical act. What he would be doing would be similar to what any one of his patients do when they 'take care of themselves': They do not perform a medical act *because they are not doctors*.

To understand the impossibility of performing a medical act on oneself, it suffices to consider that the verb to care is an example of what Descombes calls elsewhere in his book

> "the sociological verbs". These verbs have the following grammatical trait: they do not have a reflexive form. When these verbs are constructed as a reflexive verb, their meaning changes. The reason for this semantic mutation is that these verbs express social concepts, concepts that can only be applied to the social man, not to the natural Man as described by artificialist theories (Descombes 2004, 311).

Inasmuch as medical activity is a social activity, it follows *that the physician Callias cannot take care of himself* as he would take care of his other patients. Similarly, when physicians attempt to transform patients into their own physicians, *they also cannot 'take care of themselves'* in the same sense as when 'the physician takes care of the patient.'

But if the care that Callias offers himself is not a medical activity, then what is it? It is much closer to what is called 'caring about oneself' meaning to care about one's existence (Descombes 2004, 230). Descombes's analyses also (p. 254), in connection to a text by Michel Foucault concerning the care of self, the Socratic injunction to Alcibiades to take care of himself, something one cannot leave to others (Descombes 2004, 254). The conclusion that the patient cannot take care of herself in the sense of a medical activity, as presented here, obviously does not mean that she does not often try to do it by applying what she has read in a medical dictionary and especially on the Internet. We may nonetheless wonder if this desire is not an illusion, due precisely to the sociological character of the verb care. We will have the opportunity in the conclusion of this book to come back to the generalization of the notion of care that emerges from this analysis, leading to a better understanding of the meaning of the expression 'to take care of oneself', one of the keys to our entire investigation.

8.4.7 Theoretical Limits of Empowerment

Attempting to make patients medically responsible runs up against both psychological as well as semantic limits. Moreover, radical empowerment of people to administer their own treatment has its own perils, if *it is not carefully applied*. Some patients, due to their life circumstances or to *innate* preference for the present, *simply cannot* engage in long-term health actions. But sometimes it is crucial, even in the case of chronic diseases, to begin an aggressive treatment. Thus, there is no sweeping answer. To content oneself with giving patients information or to ask what their preferences are without stating one's own recommendations would be an overly simplistic—one might even say cowardly—view of patient education. Such a clinical approach ignores the psychological reality whose philosophical interpretation we have outlined here.

In sum, to decide which model of the therapeutic relationship is the most appropriate, in view of these thoughts on patient autonomy, one must first consider the fluidity of this autonomy: Is it not conceivable that even the most autonomous patient, at certain times, may ask to *rely* on the physician's decision? Obviously medical art consists in being able to choose, at the right moment, between the different models of the therapeutic relationship. But then, due to the diversity of patients, there will always be some who will not want the responsibility. A study of patients suffering from cardiac disease showed the heterogeneity of patient behavior in regards to their desire to participate in the medical decision, and the authors simply recommend asking the patient what her wishes are, rather than using a particular model (Clarke et al. 2004).

Thus, as we have shown here, it is possible that for certain patients, or at certain times, the practice of autonomy is simply not possible. It appears that a certain dose of paternalism could, at least in certain cases or from time to time, be adopted as a legitimate step—though today we would talk about neo-paternalism, which calls for a discussion with the patient (Weiss 1985; Savulescu 1995) to clarify her preferences (Brody 1993). Katz (1984) has shown the importance of dialogue, a conversation, for breaking the silence between the patient and the physician which leads to the classic form of paternalism. This conception of the therapeutic relationship recognizes that the complexity of medical reasoning implies that it is impossible to separate facts from value judgments: This suggests that paternalism cannot be completely absent in medical decision making (Wulff 1995).

This also means that we cannot be content with a purely 'informative' model or even with an 'interpretative' model where the physician's only role is to help the patient elucidate her preferences, without trying to influence the patient by revealing her own. Ingelfinger suggests that to continue to deserve the name 'doctor', the physician must be capable of taking responsibility for the medical decision and not letting it fall on the patient's shoulders (Ingelfinger 1980). His Editorial was titled *Arrogance*.

Among the four models of the therapeutic relationship, our proposed model resembles the *deliberative model* most of all. Certainly, it is open to interrogation: (1) Could the deliberative model be a return to paternalism in disguise? (2) Is it appropriate for a physician to criticize the patient's preferences and to propose her own? If the idea of preferences implies the idea of values, by what right can one make such a *value judgment*, when facing an ethnic, cultural, social and emotional diversity of individuals to be treated, this diversity implying a plurality of values? (3) If the decision depends on the physician's preferences, how can we be assured that it will not depend on the physician in question, since there is a diversity of physicians, just as there is of patients? (4) What is a 'good' choice? From the physician's point of view, Osler's aphorism that 'medicine is a science based on uncertainty and an art of probabilities' remains true. And from the patient's point of view, how is she to reason using the information she has been given?

Still, the deliberative model eases the tension that arises when the principle that dictates the respect of patient autonomy contradicts the principle of beneficence, a tension which will result in nonadherence. The interpretative model may tend to give precedence to the principle of autonomy to the detriment of beneficence. And if a certain dose of paternalism is found in the deliberative model,

> its imposition maybe […] a reflection of what many patients do wish for. That is to have medical care based on mutual trust between doctor and patient with the assumption that as a member of a profession the doctor will make choices in the best interests of the patient (Shinebourne and Bush 1994, 407).

8.4.8 Respecting Patient Autonomy

So the physician will know when to stop at the limits of autonomy that the patient is ready to receive. And she can do it with dignity because respecting patient autonomy also means respecting the fact that the patient can refuse autonomy, i.e., respecting the patient manifesting her autonomy by freely choosing to renounce it—*which is also a manifestation of the reflective capacity of her mind*. Who would deny that even bound, Ulysses remains autonomous? This is perfectly expressed by Alfred Tauber:

> How is autonomy to be exercised if a patient becomes 'preoccupied' and unable to take part in making medical decisions? Though such persons run the risks associated with paternalism […] Patients delegate such decisions, because they reasonably believe that physicians are more capable of assessing technical choices for them (Tauber 2005, 142).

The therapeutic alliance is the best we can do. It implies not only the sharing of beliefs and desires but also the possibility that each may impose constraints on the other, if there is trust between the two, Trust is not a simple word, and the importance of this concept for the therapeutic relationship will be analyzed in the next chapter.

8.4.9 Necessary Coexistence of Two Medical Models

The goal of this chapter was to show that there is another, ethical, aspect of patient adherence/non adherence, seen as her response to the conflict between the principles of beneficence and autonomy. This dialectical tension can only be resolved by the cohabitation of two medical models of health. First is the biomedical model, which is centered on a curative approach, with diagnostic investigations and therapeutic acts conducted by the physician alone. And second is the global model, which includes at the outset of an illness a complex group of organic, psychosocial and environmental factors, and which secures the collaboration of all health care professionals in the fight against the disease, as well as the active participation of the patient (Harkness 2005). The psychosocial model of medicine was first proposed by George Engel (1977).

It is too often tempting for the physician to see in the person who enters her office a patient, not to say a disease. Her medical studies have trained her to dissect medicine into diseases, syndromes and symptoms, and into treatments for these diseases, syndromes and symptoms, running the risk of forgetting that it is not a 'patient' who comes to see her, but a *person* who has a *history* of which the illness is only a part, a history that the patient *herself* can tell the physician.

Descombes (2004, 135), reminds us of Ricœur's (1995) definition of a person in *Oneself as Another*: "A person is at the same time someone about whom we can talk as an individual with a personality, and someone *who can speak*, and who can even speak about oneself, in the first person." This is the basis of narrative medicine, of the ethical importance of *conversation* in medicine (Brody 1993, 207–215). As Tauber shows in his book on patient autonomy, the education of physicians takes place in a biomedical universe constructed of facts having more and more to do with cold numbers and devoid of emotions, and it is time to fight this tendency and show future doctors the importance of empathy in medical actions. This consists in viewing the patient as a person, attempting to know her mental states and preferences, coming to know *precisely what makes her a person*.

But this does not mean that empathy must be substituted for medical responsibility. Because the patient comes to see a physician *who is also a person with her own mental states*, among them knowledge, competence, beliefs, desires, and the emotions that make her capable of empathy. What the patient expects is for the physician-person to be capable of making the best decisions for the patient-person. And when the patient-person sometimes goes so far as to ask the physician-person to force her to take care of herself, perhaps she does it because she hopes that this will be done with the cold objectivity of which she knows herself to be incapable. The global psychosocial model must not replace the biomedical model: The two models must coexist.

8.4.10 Training in Autonomy: For a Medicine of the Person

We saw that in the deliberative model, autonomy was likened to the development of the person. This is not surprising if we consider the concepts elaborated in the two founding texts we have quoted, Frankfurt's—the reflective activity of the mind defines the person, and Dworkin's—the reflective activity of the mind defines autonomy. Transitively, does autonomy define the person? And, inversely, does the notion of person not always imply at least a *potential for autonomy*? The concept of autonomy acquires then a dynamic significance: *It is always a potential.*

This is why a psychological discourse stating that therapeutic autonomy is not always exercised must not be an excuse for giving up on the Kantian ideal; we must still try to *foster* therapeutic autonomy. We must never forget that autonomy is always *possible*, precisely because the patient is a person. The development of this *potential*, i.e., the promotion of the autonomy of the patient, is one of medicine's duties, within the limits that were listed earlier, and it is particularly one of the essential goals of patient education. Patient education, in the largest sense of the word, appears to be the means to solving the paradox at the heart of this chapter.

Finally, in regard to autonomy, we have come back to respect for the person. Respect for the person includes respect for therapeutic autonomy. By entering a therapeutic situation, the person became a patient. By respecting the person, whether she is autonomous or not, or whether she is autonomous but freely renounces her autonomy for a time, the physician restores the status of person to the patient.

In the next chapter, we will examine how doctors may fail to consider the future interests of their patients: We will suggest that, like patients' nonadherence to medical recommendations, doctors' clinical inertia could represent a case of clinical myopia. The analysis of patient *nonadherence* led to a better understanding of *adherence*; in the same vein, investigating the mechanisms of clinical inertia will provide a clue making it possible to revisit the meaning of care in chronic diseases, of patient education, and of trust in the patient-physician relationship.

References

Adams RJ, Smith BJ, Ruffin RE. Patient preferences for autonomy in decision making in asthma management. Thorax. 2001;56:126–32.

Anderson RM, Funnell MM, Carlson A, Saleh-Stattin N, Cradock S, Skinner TC. Facilitating self-care through empowerment, a new paradigm. In: Snoek FK, Skinner TC, editors. Psychology in diabetes care. Chichester: Wiley; 2000.

Balint M. Le Médecin, son malade et la maladie. Bibliothèque scientifique Payot; 1988 (The doctor, his patient, and the illness; 1957).

Barrier PA, Li JT, Jensen NM. Two words to improve physician-patient communication: what else? Mayo Clin Proc. 2003;78:211–4.

Beauchamp T, Childress J. Principles of biomedical ethics. Oxford: Oxford University Press; 2001.

Bloch-Lainé JF. http://psydocfr.broca.inserm.fr/conf&rm/conf/conftox/Bloch.html. Access 5 May 2013.

Brody H. The four principles and narrative ethics. In: Gillon R, Lloyd A, editors. Principles of health care ethics. New York: Willey; 1993.

Canguilhem G. The normal and the pathological (trans: Fawcett C). New York: Zone Books; 1989.

Canguilhem G. Écrits sur la médecine. Éditions du Seuil; 2002.

Clarke G, Hall RT, Rosencrance G. Physician-patient relations: no more models. Am J Bioeth. 2004;4:W16–9.

Davidson D. First person authority. In: Subjective, intersubjective, objective. Oxford: Oxford University Press; 2001.

Davidson D. Paradoxes of irrationality. In: Problems of rationality. Oxford: Clarendon Press; 2004.

Delpla, I. Quine, Davidson, le principe de charité. P. U. F., Collection Philosophies; 2001.

Descombes V. L'Inconscient malgré lui. Éditions de Minuit; 1977.

Descombes V. The mind's provisions (trans: Schwartz SA). Princeton: Princeton University Press; 2001.

Descombes V. Le Complément de sujet, enquête sur le fait d'agir de soi-même. nrf essais, Gallimard; 2004.

Doherty C, Doherty W. Patients' preferences for involvement in clinical decision-making within secondary care and the factors that influence their preferences. J Nurs Manag. 2005;13:119–27.

Dworkin G. The theory and practice of autonomy. Cambridge: Cambridge University Press; 1988.

Emanuel EJ, Emanuel LL. Four models of the physician-patent relationship. JAMA. 1992;267:221–6.

Elster J. Ulysses unbound. Cambridge: Cambridge University Press; 2000.

Ende J, Kazis L, Ash A, Moskozitz MA. Measuring patients' desire for autonomy: decision making and information-seeking preferences among medical patients. J Gen Int Med. 1989;4:23–30.

Ende J, Kazis L, Moskozitz MA. Preferences for autonomy when patients are physicians. J Gen Int Med. 1990;5:506–9.

Engel G. The need for a new medical model: a challenge for biomedicine. Science. 1977;196:129–136.

Engel P. Perspectives sur Davidson. In: Lire Davidson. L'Éclat; 1994.

Engel P. Les croyances. In: Notions de philosophie (under the direction of D. Kamboucher). Editions folio, Gallimard; 1995.

Feldman SR, Chen GJ, Hu JY, Fleischer AB. Effects of systematic asymmetric discounting on physician-patient interactions: a theoretical framework to explain poor compliance with lifestyle counseling. BMC Med Inform Decis Mak. 2002;2:8.

Frankfurt H. The importance of what we care about. Cambridge: Cambridge University Press; 1988.

Funnell MM, Anderson RM, Arnold MS, Barr PA, Donnelly M, Johnson PD, Taylor-Moon D, White NH. Empowerment: an idea whose time has come in diabetes education. Diab Educ. 1991;17:37–41.

Harkness J. Patient involvement: a vital principle for patient-centered health care. World Hosp Health Serv. 2005;41:12–6.

Hippocrates. Hippocratic writings (trans: Lloyd GER, Chadwick J). Penguin Classics; 1978.

Ingelfinger FJ. Arrogance. N Engl J Med. 1980;303:1507–11.

Kahneman D. Thinking, fast and slow. Westminster: Allen Lane, Penguin Books; 2011.

Kahneman D, Tversky A. Prospect theory: an analysis of decision under risk. Econometrica. 1979;47:263–92.

Katz J. The silent world of doctor and patient. London: Collier, MacMillan Publishers; 1984.

Lee SJ. Putting shared decision making in practice. In: Enhancing physician-patient communication. American Society of Hematology; 2002.

Levinson W, Kao A, Kuby A, Thisted RA. Not all patients want to participate in decision making. A national study of public preferences. J Gen Int Med. 2005;20:531–5.

Lewis D. Dispositional theories of values. Proc Aristotelian Soc. 1989;63:113–37.

Livet P. Émotions et rationalité morale. P. U. F., Collection Sociologies; 2002.

Mansell D, Poses RM, Kazis L, Duefield CA. Clinical factors that influence patients' desire for participation in decisions about illness. Arch Int Med. 2000;160:2991–6.

Marchand C, d'Ivernois JF, Assal JP, Slama G, Hivon R. An analysis, using concept mapping, of diabetic patient's knowledge, before and after patient education. Med Teacher. 2002;24:1–99.

Moro MR. Enfants d'ici venus d'ailleurs. La Découverte; 2002.

Prochaska J. In: Prochaska J, Norcross JC. Person-centered therapy. In: Systems of psychotherapy, a transtheoretical analysis. Pacific Grove (CA): Brooks/Cole; 1994.

Reach G. Obstacles to patient education in chronic diseases: a transtheoretical analysis. Patient Educ Counsel. 2009a;77:192–6.

Reach G. Linguistic barriers in diabetes care. Diabetologia. 2009b;52:1461–63.

Reach G, Zerrouki A, Leclerc D, d'Ivernois JF. Adjustment of insulin doses: from knowledge to decision. Patient Educ Counsel. 2005;56:98–103.

Rheault WL. Incorporating locus of control theories into patient education practices. Physiother Can. 1982;34:152–6.

Ricœur P. Oneself as another (trans: Blamey K). Chicago: University of Chicago Press; 1995.

Rogers CR. A theory of therapy, personality and interpersonal relationships, as developed in the client-centered framework. In: Koch S, editor. Psychology: a study of science. vol. 3. New York: Mc Graw Hill; 1959.

Savulescu J. Rational non-interventional paternalism: why doctors ought to make judgments of what is best for their patients. J Med Ethics. 1995;21:327–31.

Schillinger D, Piette J, Grumbach K, Wang F, Wilson C, Daher C, Leong-Grotz K, Castro C, Bindman AB. Closing the loop: physician communication with diabetic patients who have low health literacy. Arch Int Med. 2003;163:83–90.

Shinebourne EA, Bush A. For paternalism in the doctor-patient relationship. In: Gillon R, Lloyd A, editors. Principles of health care ethics. New York: Wiley; 1994.

Strull WM, Lo B, Charles G. Do patients want to participate in medical decision making? JAMA. 1984;7(252):2990–4.

Tauber AI. Patient autonomy and the ethics of responsibility. Cambridge: MIT Press; 2005.

Valéry P. Colloques. Socrate et son Médecin. Paris: Gallimard; 1955.

Weiss GB. Paternalism modernized. J Med Ethics. 1985;11:184–7.

World Health Organization Report. Sabaté E, editor. Adherence to long-term therapies, evidence for action. Geneva, Switzerland; 2003.

Wulff HR. The inherent paternalism in clinical practice. J Med Philos. 1995;20:299–311.

Chapter 9
Doctors' Clinical Inertia as Myopia

Abstract The inadequate physician adherence to current good practice guidelines has been recently described as clinical inertia. The aim of this chapter is to show that clinical inertia shares with patients' nonadherence to medical prescriptions the appearance of myopia, and may be, at least sometimes, the result of inappropriate use of empathy by the doctor. I propose that clinical inertia occurs when the doctor imagines her patient's feelings, for instance that she will refuse the treatment, and becomes overly involved in the immediacy of those emotions at the expense of the future. The doctor, in exercising empathy, should not forget that her own preference is different, and that she should propose a treatment which would protect her patient's future: Not doing it is clinical inertia. Finally, how, in the context of the autonomy principle, someone (a Health Care Provider) can decide what is good (a treatment) for someone else (a patient) without falling into paternalism? Actually this analysis leads to a paradox: not only is the principle of benevolence sometime conflicting with the principle of autonomy, but physician's benevolence may enter in conflict with the mere respect of the patient; I propose a solution to this paradox relying on the importance of patient education and trust in the patient-physician relationship.

As shown in this book, the efficiency of medical care is often jeopardized by the lack of patient adherence to recommendations. But there may be also a lack of prescription by the doctor. While there are in several fields of medicine official guidelines to help physicians decide the most appropriate therapy, it happens that such guidelines are not followed by doctors, as they should. This inadequate physician *adherence to current guidelines* has been recently described as *clinical inertia* (Phillips et al. 2001).

In this chapter I intend to show that patients' nonadherence to medical prescriptions and physicians' nonadherence to guidelines may represent homologous phenomena, having a similar explanation: A failure to give precedence to the long-term benefits of treatment is a common driving force for patient nonadherence and physician clinical inertia. In cases of clinical inertia, the doctor, despite knowing what ought to be done clinically, does not do it. The doctor seems unable to act for the future interests of her patient. Again, how is this *possible*?

© Springer International Publishing Switzerland 2015
G. Reach, *The Mental Mechanisms of Patient Adherence to Long-Term Therapies*,
Philosophy and Medicine 118, DOI 10.1007/978-3-319-12265-6_9

The aim of the following discussion is to show that clinical inertia may be, at least sometimes, the result of inappropriate use of empathy by the doctor. Given the largely unquestioned belief in the merits of empathy in the doctor-patient relationship, it is odd that it should be—as proposed here—an occasional impediment to proper care. I will propose that clinical inertia occurs when the doctor imagines her patient's feelings, for instance that she will refuse the treatment, and becomes overly involved in the immediacy of those emotions at the expense of the future. The doctor, in exercising empathy, should not forget that her own preference is different, and that she should propose a treatment which would protect her patient's future: Not doing it is clinical inertia.

I will propose that using a new form of sympathy, defined explicitly as an emotion involving in the doctor's mind concern for both the patient's present *and* future, may represent a way to reconcile the importance of emotion in the doctor-patient relationship with the need to avoid clinical inertia. Finally, by revisiting the very concepts of the autonomous person, patient education, and trust in the patient-doctor relationship, I will show how empathy, combined with this form of future-oriented sympathy, can be successively integrated in an ethical pathway which eschews paternalism and fosters the deliberative model of the patient-doctor relationship: This leads to an ethical definition of patient education.

9.1 Clinical Inertia: Definition and Logical Description

> The goals for management are well defined, effective therapies are widely available, and practice guidelines for each of these diseases have been disseminated extensively. Despite such advances, health care providers often do not initiate or intensify therapy appropriately during visits of patients with these problems. We define such behavior as *clinical inertia*—recognition of the problem, but failure to act.

This definition of clinical inertia, given in the first article describing the phenomenon (Phillips et al. 2001), is linked with the existence of a guideline, i.e. a well-defined clinical pathway. In short, clinical practice guidelines function to help the doctor overcome clinical inertia. However, it turns out that doctors often don't follow them. This in turn suggests an epistemological link between the recognition of clinical inertia as a clinical issue and the development of a new way of practicing medicine: Evidence-based medicine leading to the publication of guidelines.

Lawrence Phillips and colleagues recognized that there are situations where a strict application of guidelines could result in over-treatment. On the other hand, the application of guidelines may be jeopardized by two other explanations: Doctors may find that guidelines are too rigid, or too removed from clinical reality (Bachimont et al. 2006), or they may give as an excuse the fact that they have neither the time nor the training to apply the guidelines, for instance, to explain lifestyle changes to their patients (Cogneau et al. 2007). More recently, the focus has shifted to three other explanations of clinical inertia: (1) Competing demands—Parchman et al. (2007) highlighted the fact that clinical inertia is more frequent when appointments are short, and, especially, that this effect is aggravated when an intercurrent problem occurs; (2) Clinical uncertainty—for instance, in the case of hypertension management, when it might not be clear that the patient's usual blood pressure is elevated and that a doctor needs to take action (Kerr et al. 2008; Turner et al. 2008; Phillips 2008); (3) Misjudgement—the doctor mistakenly decides that the guidelines do not apply to this particular patient (this happens often, for instance, in the care of elderly patients) (Miles 2010).

However, even without these limits, clinical inertia may occur. For instance, it has been shown (Drass et al. 1998) that general practitioners consider as a mean 6.9 % as the target for HbA1c in type 2 diabetes, which is very close to the recommended target (7 %), and that health care providers are able to identify 88 and 94 % of patients with good or insufficient control, respectively (el-Kebbi et al. 1999). Nevertheless, a study showed that treatment intensification occurred in only 37.4 and 45.1 % of diabetic patients with HbA1c higher than 8 %, who consulted a general practitioner or a specialist, respectively (Shah et al. 2005).

Let us propose the following definition of "strict" clinical inertia, Doctor (D) is clinically inert with patient (P) if and only if:

(a) There is a guideline G recommending prescription X
(b) The doctor is aware of the existence of guideline G
(c) The doctor has the resource to prescribe X
(d) The doctor judges that, all things considered, guideline G is pertinent for patient (P)
(e) Doctor does not prescribe X to patient (P).

Note the importance of (d): As discussed in detail in my *Clinical Inertia, A Critique of Medical Reason* (Reach 2014a), it is essential to distinguish true clinical inertia from situations where the physician's judgment recognizes that a guideline should not be applied to this patient she is faced with: This would not be clinical inertia but instead often appropriate inaction. What's more, giving priority to the physician's clinical judgment is what the theory of Evidence-Based Medicine states time and again. Here, we consider *true* clinical inertia, that which cannot be justified by the doctor.

On the other hand, there are a number of clinical situations where there is no *published* guideline, and most people would agree that the absence of a well-defined guideline should not obstruct treatment: Here, proper care means compliance with *common clinical judgment*. We may therefore also use a less strict definition of clinical inertia, wherein the doctor acts against her best judgment.

This formulation of the problem is very similar to the way philosophers define weakness of the will, or incontinent action, or *akrasia*. Let's recall that Donald Davidson described incontinent action as follows (Davidson 2001, 40).

In doing x, an agent acts incontinently if and only if:

(a) The agent does x intentionally
(b) The agent believes there is an alternative action y open to him
(c) The agent judges that, all things considered, it would be better to do y than to do x.

It is therefore tempting to consider that clinical inertia, like patient nonadherence, represents a case of *akrasia*. We saw previously that Christine Tappolet proposed a role for emotions in the genesis of akratic actions (Tappolet 2003, 97–120). First, she considered that

> emotions could have the same function with respect to values as perceptual experiences have for colors and shapes, so that we can say emotions are perception of values.

Therefore,

> actions caused by emotions can be explained in terms of the perceived value. The value that is perceived, whether correctly or not, makes the action intelligible. Suppose a bear attacks me. The fear I experience does not only save my life when it causes my running away; it also makes the action intelligible [...] This claim can be generalized to cases of akratic action [...] If this is right, emotions are not only causally involved in cases where we act against our best judgment; they make the action intelligible, even though we judge that another course of action would have been better all things considered.

We will apply this conception of *akrasia* to the case of clinical inertia and propose that clinical inertia, this "failure to act" according to the definition given by Phillips, can be caused and even explained by some emotions felt by a doctor when she has to initiate or intensify a treatment, whether the values that these emotions express are correctly perceived or not.

9.2 Empirical Evidence: The Paradigm Case of Psychological Insulin Resistance

Consider the phenomenon of "psychological insulin resistance" initially described by Leslie (Leslie et al. 1994). Psychological insulin resistance refers to the delayed start in using insulin, typically in patients with type 2 diabetes who, gradually, come to require insulin. Here insulin resistance is not physiological, but psychological. It consists both of a resistance to accepting insulin among patients and a resistance to prescribing insulin among doctors. In this latter case, it can be viewed as a special case of clinical inertia. The research on psychological insulin resistance suggests doctors are swayed not only by their beliefs (e.g., insulin will not work) but also by their emotions, such as the fear of causing hypoglycemia and weight gain. Patrick Phillips pointed out the special significance is the doctor's fear of the patient's reaction to bad news (Phillips 2005).

The doctor may think empathetically: My patient will be afraid of the injections, she will lose hope that she can manage her illness, or perhaps she will see me and my treatment as a failure; therefore, the doctor thinks, she will refuse to start insulin. The doctor holds off prescribing insulin, pushing the difficult moment off to the future. In so doing, the doctor does not make the appropriate treatment recommendation or even begin working towards it; instead, the doctor closes off that avenue indefinitely. The doctor is convinced a priori that the patient will refuse. On the other hand, if the doctor had recommended insulin and then gave up after the patient refused, it would not be a case of clinical inertia but simply an expression of the respect of patient's autonomy.

We can therefore propose the following mechanism leading to the doctor's clinical inertia (Fig. 9.1).

In the patient's mind, there is, as shown in this book, a conflict between the immediate concern of starting insulin and the long-term consideration of the benefit of accepting insulin treatment to preserve her health. The patient will manifest psychological insulin resistance if the focus is on the short term concern. In the doctor's mind, there is also a conflict between the immediate empathetic fear that the patient will refuse insulin and the intention to follow current guidelines indicating that insulin must be prescribed to preserve the patient's health over the long-term. Clinical inertia can therefore be seen, at least in part, as a victory of the doctor's empathetic consideration of the patient's immediate concern over the professional duty to follow medical guidelines. Another example of empathetic interaction between minds of a doctor and a patient is represented by the observation that doctors are more often clinically inert (do not intensify therapy) with patients who are nonadherent to antidiabetic agents (Grant et al. 2007). At this point in the discussion, it appears—somewhat surprisingly—that the doctor's empathy is involved in her clinical inertia.

Fig. 9.1 Clinical inertia as myopia. From Reach (2014a)

9.3 Empathy and Sympathy

Remember the classic definition formulated by Rogers (1959), empathy is defined as the ability

> to perceive the internal frame of reference of another with accuracy and with the emotional components and meanings which pertain thereto as if one were the person, but without ever losing the '*as if*' condition. Thus, it means to sense the hurt or the pleasure of another as he senses it and to perceive the causes thereof as he perceives them, but without ever losing the recognition that it is as if I were hurt or pleased and so forth.

The "as if" condition is important: If this condition is absent, feeling another person's emotions is not empathy but sympathy (i.e., emotional identification). The distinction between empathy and sympathy was clarified by Wispé (1986) as follows:

> To know what it would be like *if* I were the other person is empathy. To know what it would be like *to be* that other person is sympathy. In empathy I act '*as if*' I were the other person. In sympathy *I am* the other person. The object of empathy is to 'understand' the other person. The object of sympathy is the other person's 'well-being.'

This explains why empathy, and not sympathy, has been privileged in the patient-doctor relationship. According to Wispé, sympathy is not the mode for therapeutic interaction:

> Sympathy does not facilitate accurate assessments. One cannot be sympathetic and objective. Sympathy lends itself to emotional distortions. Sympathy can lead to closer emotional identification and to peremptory rescue actions in the patient's behalf [...] Compassionate understanding is one thing in therapy; sympathy is another.

Similarly, Mohammadreza Hojat proposed that the relationship with patient outcomes is positive and linear for empathy but curvilinear (having an inverted U-shape) for sympathy. In other words, only a small dose of sympathy may be beneficial (Hojat 2007).

9.4 The Paradox of Empathy in Medical Care

Clearly, if one refers to these definitions of empathy and sympathy, we can conclude that both of them can lead to clinical inertia. Let us say that a Doctor D sees a patient, Ms P who is psychologically resistant to the idea of starting insulin, giving priority to her fear of insulin and refusing to accept the fact that insulin may preserve her health. Doctor D considers empathetically Ms P's concern. Even without forgetting the "as if" condition, Doctor D understands Ms P's fear about insulin and acts accordingly: She does not prescribe insulin and will therefore be clinically inert. If, on the other hand, Doctor D forgets the "as if condition," she expresses sympathy for Ms P (emotional identification) and, of course, does not prescribe insulin either, which also leads to clinical inertia.

We therefore arrive at a paradox of empathy in medical care: While empathy is presented as a cornerstone in the doctor-patient relationship, Doctor D will actually prescribe insulin to Ms P (i.e., will not be clinically inert) if (i) she does not express empathy at all (paternalism) or (ii) if she behaves empathetically, but with a definition of the "as if" condition of empathy which implies that she remembers that she is a doctor, and that to be a doctor means giving priority to the patient's future well-being and *not to the patient's own concern*. In other words, clinical inertia can be avoided if the doctor either does not express empathy or expresses empathy but does not act accordingly, which is not impossible: Philosopher Stephen Darwall (2002), using the example of a child on the verge of falling into a well, remarked that "empathy consists in feeling what one imagines he feels, or perhaps should feel (fear, say), or in some imagined copy of these feelings, whether one comes thereby to be concerned for the child—or not." He added: "Empathy can be consistent with the indifference of pure observation or even the cruelty of sadism. It all depends on why one is interested in the other's perspective." Of course, coming from a doctor, such behavior would be recognized as an aberrant response, and probably as pathological.

9.5 Another Conception of Sympathy

Darwall, in his book, *Welfare and Rational Care*, proposed another definition of sympathy, which may be relevant in the context of the psychological relationship between doctor and patient. He calls sympathetic concern or sympathy

> a feeling or emotion that (i) responds to some apparent obstacle to an individual's welfare, (ii) has that individual himself as object, and (iii) involves concern for him, and thus for his welfare, for his sake. Seeing a child on the verge of falling [into a well], one is concerned for his safety, not just for its (his safety's) sake, but for *his* sake. One is concerned for *him*. Sympathy for the child is a way of caring for (and about) him (Darwall 2002, 50–72).

Sympathy, in this Darwallian sense, is clearly different from empathy:

> empathy is the imaginative occupying of another's point of view, seeing and feeling things as we imagine her to see and feel them. Sympathy for someone, on the other hand, is felt not as from her standpoint but as from the perspective of someone (anyone) caring for her. Empathizing with someone in a deep depression, we imagine how things feel to her, for example, how worthless she feels. When, however, we view her situation with sympathy (a sympathy she perhaps cannot muster for herself), she and her welfare seem important, not worthless.

Similarly, psychoanalyst David Black observed that there are two senses of the word sympathy. First, sympathy refers to "a spontaneous capacity to be directly affected by the feeling state of others". Let us say that this form of sympathy is "*a capacity*", like empathy, and it is by and large empathy from which the "as if" condition was removed. It represents the sympathy in the sense of Wispé (emotional identification). Second, sympathy can refer to another concept, according to Black:

> a warm concern for the feelings of others. Sympathy in this sense, also called compassion, is an emotion, or a range of emotions, akin to sorrow and belonging with the depressive position group, and like other emotions, it can be highly developed, repressed, split off, etc. (Black 2004).

Here, sympathy is *an emotion*, and this meaning of the word sympathy represents actually the sympathy in the sense of Darwall, who described it also as an emotion.

9.6 Care, Sympathy, Beneficence, and Love

The word "care" comes from old English word, caru, éearu: care, concern, anxiety, sorrow, grief, trouble; this old English word was derived from the proto-Germanic word, *karō: care, sorrow, cry; that proto-Germanic word, in turn, came from the proto-Indo-European word, *ǵār-, *gÀr-, voice, exclamation. Thus, etymologically at least, care seems to be the emotional answer to a cry: Is not care the innate answer of the mother, when she hears, for the first time, the cry of her child? Therefore, the words care, sympathy, and beneficence may have a synonymous meaning.

One thinks also to the Aristotelian concept of *philia*, one of the Greek words for love, often translated as friendly feeling:

> We may describe friendly feeling towards any one as wishing for him what you believe to be good things, not for your own sake but for his, and being inclined, so far as you can, to bring these things about. (Aristotle 1381)

9.7 Care as a Special Form of Sympathy

However, something more is needed. Take again Doctor D and Ms P. Even if Doctor D exercises this form of sympathy, this does not mean that she will prescribe insulin. Indeed, she may consider that Ms P's fear of insulin represents something that jeopardizes her welfare. Caring for her patient, Doctor D may want to improve her welfare and decide not to prescribe insulin. Again, she manifests clinical inertia.

However, Doctor D may also manifest this kind of sympathy for her patient by defining her welfare, but not just *any* welfare: Her *future* welfare. For instance, having to decide between the patient's immediate feelings (the fear of insulin) and what she, as a doctor, considers to be the patient's future welfare (what is enshrined in evidence-based guidelines), she would decide in favor of the guidelines, thus avoiding the pitfall of clinical inertia.

We propose therefore that to avoid clinical inertia, the doctor should practice not only empathy, appreciating the feelings of her patient, but also a new form of sympathy, defined as an emotion that takes the three criteria defined by Darwall: "(i) Responds to some apparent obstacle to an individual's welfare, (ii) has that individual himself as object, (iii) involves concern for him, and thus for his welfare," and adds a fourth condition, (iv) specifying clearly that the emotion involves concern for the patient's future.

This conclusion (that doctors are concerned with the future of their patients) may be considered as a truism. It takes its real meaning here, at the end of a book aimed to explain patients' adherence: If clinical inertia, at least in some cases, is due to the doctor failing to consider the future of her patient, in that case, doctor's clinical inertia shares with patient's nonadherence the fact to be a case of "clinical myopia" (Reach 2008).

In Darwall's definition, sympathy is "an emotion which has for its object the person herself, involving concern for her, and thus for her welfare, for her sake". Including this fourth condition in the definition of sympathy entails accepting that, in the doctor-patient relationship, the concept of a person encompasses the idea of the future. Interestingly, an emphasis on prognosis had been recommended by Hippocrates (Hippocrates 1978):

> It seems to be highly desirable that a physician should pay much attention to prognosis. If he is able to tell his patients when he visits them not only about their past and present symptoms, but also to tell them what is going to happen, as well as to fill in the details they have omitted, he will increase his reputation as a medical practitioner and people will have no qualms in putting themselves under his care. Moreover, he will the better be able to effect a cure if he can foretell, from the present symptoms, the future course of the disease (Hippocrates, Prognosis).

9.8 The Respective Values of Immediacy and Future

When I introduced the hypothesis of a role for certain emotions in the phenomenon of clinical inertia, I quoted the suggestion by Tappolet (2003) that emotions are the perception of values. In Darwall's definition, the sympathy that the doctor feels for the patient is the emotion that has "the patient herself" for object. The doctor sympathizing with the patient has to decide between the value of the patient's emotions which are often essentially present-oriented, and the value of the person herself (i.e., her future). According to the Construal Level Theory developed by Trope and Liberman (2003), concepts related to the future are categorized in our mind as having a higher level than immediate ones. However, they are also more abstract. This may explain why they are difficult to apply, both for the patient and for the physician, and why patient nonadherence and doctor clinical inertia are so frequent.

Harry Frankfurt, in his essay *"The importance of what we care about"* (Frankfurt 1988), observed "that the outlook of a person who cares about something is

inherently prospective; that is, he necessarily considers himself as having a future."
The interesting point here is that if we transpose this definition of care to health
care, we are in a situation where the doctor who takes care of the patient considers
not only herself (the subject who cares) but also her patient (the object of care) as
having a future.

9.9 Empathy, Sympathy, and the Ethical Dynamics
of the Patient-Doctor Relationship

The key point in Darwall's argument is that a person's well-being is defined not
only by the person herself but also by "someone (anyone) caring for her". Anyone
caring for her, *because* she cares for her, is entitled to help shape the contours
of what is good for her, in order to ensure her welfare. Of course, as already
mentioned, the patient can reject the doctor's definition of what is good, in favor
of her own definition.

Let's come back to the four models of the doctor-patient relationship proposed by Emanuel and Emanuel: (i) the *paternalistic model* where the doctor
decides for the patient, (ii) the *informative model* where the doctor gives information to the patient, (iii) the *interpretative model*, in which the goal of the
therapeutic relationship is to help the patient elucidate her preferences, as they
may not be explicit, and then to help her act by explaining which therapeutic
option is the most appropriate, taking into account the preferences expressed
during the therapeutic interaction (significantly, the physician does not influence the revelation of the preference), and (iv) the *deliberative model*, in which
the goal of the physician-patient relationship is to help the patient determine
and choose what seems best for her health. Here, contrary to the interpretative model, the physician indicates which options she finds preferable for the
patient's health (Emanuel and Emanuel 1992).

I suggest that these four models can be described as an ethical pathway.
First, introducing empathy in her relationship with the patient, the doctor proceeds from the informative to the interpretative model: indeed empathy, the
aim of which is to understand the patient, is the very attitude which will help
the patient elucidate her own preferences. Secondly, I suggest that sympathy,
according to our four-point definition, is the hallmark of the deliberative model.
Here the doctor expresses her own preferences, after having heard the patient's,
and doctor and patient move forward to craft a treatment plan. Although the
deliberative model calls for the doctor to express her preference as in paternalism, it differs from paternalism by the fact that the expression of the doctor's
preference occurs at the end of a process, described here as an "ethical pathway" shown on Fig. 9.2. Therefore, I propose that clinical inertia can be seen, at
least sometimes, as an unintended consequence of empathy. This is not to deny
that empathy is essential in the doctor-patient relationship. Certainly, it is a way
to practice the biopsychosocial model of medicine (Engel 1977), which aims

to counterbalance the coldness of a purely factual medicine. Indeed, applying empathy in the doctor-patient relationship represents a transition from the second (informative) to the third (interpretative) model. However, remaining at the interpretative stage may be dangerous: As shown herein, when the "as if" condition of empathy is misused (taking into consideration only the immediate interest of the patient) or missed (emotional identification—a primitive form of sympathy, where the doctor takes for herself the patient's immediate concern), there is a risk of clinical inertia (Fig. 9.2). This risk may be avoided if the doctor proceeds to the fourth, deliberative, model, expressing her own preferences, i.e. sympathy as it is defined herein.

Figure 9.2 represents the four models of the doctor-patient relationship as a pathway with the respective places of empathy and sympathy (as defined above). Interestingly, this is reminiscent of Darwall's description of the evolution of empathy (what he calls protosympathetic empathy), and sympathy:

> Consider the difference between the instructions: (i) imagine what someone would feel if he were to lose he only child, and (ii) imagine what it would be like for that person to feel that way.

In step (i), I share his grief projectively (empathy):

> my focus is on the child who was lost, not on the person whose grief I share. In step (ii), I turn my attention to what it must be like to live with this loss, I focus on the person himself and the ways his grief pervades and affects his life. Before, my thought was: What a terrible thing: a precious child is lost. Now my thought is: What a terrible thing for him–he has lost his precious child (Darwall 2002, 63–64).

Darwall calls state (ii) "protosympathetic empathy". Indeed it is not yet sympathy and does not necessarily give rise to sympathy. Sympathy occurs if I decide that the situation of this father represents an obstacle to his welfare, and if, (iii), I have the intention to relieve his suffering.

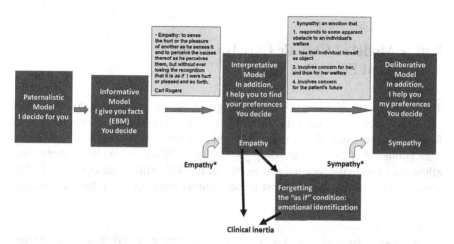

Fig. 9.2 The ethical pathway of patient-doctor relationship

In this chapter I have proposed a relationship between clinical inertia and empathy. Clearly, the case for this hypothesis was not made empirically, and there are other explanations for clinical inertia [see my *Clinical Inertia, a Critique of Medical Reason* (Reach 2014a)].

Nevertheless, I am highlighting here a paradoxical effect of empathy in medical care: Exercising empathy is necessary if one wishes to leave paternalistic behavior behind or if one eschews the pure informative model, but its exercise may present some dangers. Separating empathic ability (feeling other's emotions) and sound clinical decision-making is important for both clinicians and ethicists. Here, we propose the intervention of a new form of sympathy defined, not as *an ability*, but as *an emotion* experienced by the doctor to the perception of an obstacle to the patient's *future* welfare. This positive emotion *of her own* may help the doctor to avoid clinical inertia.

The argument presented herein would therefore fully rehabilitate the role of emotions in the patient-doctor relationship. In a paper focussing on the care of "hateful patients", James Groves remarked that

> Emotional reactions to patients cannot simply be wished away, nor is it good medicine to pretend that they do not exist (Groves 1978).

An important focus for avoiding clinical inertia may therefore be greater emphasis on interpersonal skills in the patient-doctor interaction. The deliberative model of the doctor-patient relationship, compared with the other three, requires a significantly higher degree of physician self-awareness/self-disclosure, communication and negotiation with the patient, tolerance of disagreement and patient affect, and mastery of other psychodynamic skills. Developing the ability to cope with her own emotions, i.e. counterbalancing those derived from her empathic approach of the patient with her sympathy—an emotion defined herein as her concern for her patient future welfare—becomes an essential part of the training of future doctors: Indeed, emotions matter, in the mind of patients, and of doctors as well.

9.10 A Model of Chronic Care Involving Patient Education and Trust

The three inventions[1] of evidence-based medicine, patient education and patient autonomy were contemporary in biomedicine: Evidence-based medicine was founded on the basis of Cochrane reflexion (Cochrane 1979), the *Four principles* were published in 1979 (Beauchamp and Childress 2001), and the first paper on the efficiency of patient education was published in 1972 (Miller and Goldstein 1972). These inventions may seem at first glance contradictory, since on the one hand the

---

[1] I use purposely the word "invention". In the same vein, Schneewind (1998) described the conceptualization of autonomy as an invention.

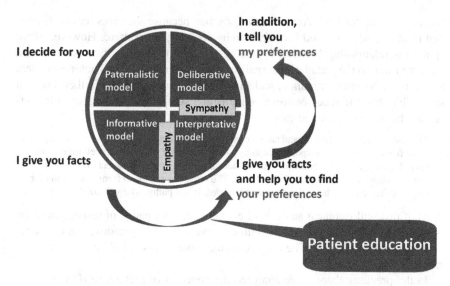

Fig. 9.3 A model of chronic care involving patient education. Originally published in (Reach 2013) © Reach 2013

doctor may wish to apply the principle of beneficence and teach the patient how to benefit from the medical progress evaluated in the framework of evidence-based medicine, and on the other hand has to respect the patient's autonomy. In the same period, the concept of patient nonadherence (formerly called noncompliance, this passive word illustrating the absence by that time of concern from the medical community for autonomy!) emerged as a major problem jeopardizing the efficiency of care (Haynes et al. 1979).

In two recent articles, I suggested that paving the ethical pathway that links the three models, informative, interpretative, and deliberative of patient-physician interaction is precisely the role of patient education (Reach 2013, 2014b).[2] I proposed a model of chronic care involving patient education (Fig. 9.3). Like in Fig. 8.3, I purposely represented the four models in a circular form. Indeed, one may fear that the deliberative model is actually nothing but disguised paternalism: Do they not have the same goal, that is, to lead the patient to do what the physicians want her to do? Is there an antinomy between care and autonomy?

If the physician only obeys to the principle of beneficence, an attitude referred to herein as sympathy according to Darwall's definition, this behavior can lead to the paradoxical conclusion that care becomes antinomic with the respect of patient's autonomy, and even with the mere patient's respect.

Indeed, according to this definition of sympathy, an individual's welfare, that is, what is important for her, is defined not by this person herself but by "someone

---

[2] What follows is taken from Reach (2013).

(anyone) caring for her". Anyone caring for her, because she cares for her, is entitled to define what is good for her, thereby ensuring her welfare. However, if we apply this relationship between sympathy and care to the specific case of health care, we run into the pitfall of paternalism. Paradoxically, care and autonomy seem therefore to become antinomic, and this antinomy may even go further. Darwall shows that this difference between empathy and sympathy is analogous to the difference between respect and care:

> The contrast between respect and care reconfirms the distinction between what is or seems good from someone's viewpoint [empathy] and what is for his good or welfare [sympathy]. Treating another's point of view as normative is a form of respect. Taking a person's welfare as normative is a form of care. The respect we demand from others calls for empathy. The care we hope for, from some at least, is sympathy. (Darwall 2002)

Thus, if this philosophical analysis of empathy and sympathy, of respect and care, is applied to the medical field of health care, we arrive at a paradox: Not only does care and autonomy seem to be antinomic; the same seems to hold true for care and respect.

In the previous chapter, we analyzed the concepts of preference (Lewis 1989), person (Frankfurt 1988), and autonomy (Dworkin 1988) and their link to the notion of reflectivity. This analysis makes it possible to solve the paradox: The apparent antinomy between care and autonomy or respect disappears if and only if I consider that when I, as a HCP, give my preferences, I do not speak to a "patient"; rather, I speak to an autonomous person, i.e., someone who is able to deliberate, and to change her mind: I do it at the end of the ethical pathway, which goes from the informative to the interpretative model (role of empathy, helping the patient to elucidate her references), and next to the deliberative model, i.e., sympathy introducing the HCP's preferences (Fig. 9.3). Under this condition, care is therefore neither paternalism nor manipulation: The deliberative model differs from the paternalistic model because it can be reached only at the end of this pathway. I propose that a major role of patient education is to pave this ethical pathway, which makes the difference between the deliberative, and the paternalistic, model. This role of patient education may well represent its very definition.

Incidentally, we already mentioned our empirical study in obese type 2 diabetic patients, showing the existence of an association between patient adherence to medication and the fact that patients declared that they fasten their seatbelt when they are seated in the back of a car, and we interpreted this association as a room for obedience in patient adherence. We proposed that a role for patient education was to replace this passive reason to be adherent to therapy by an active conscious choice (Reach 2011).

Indeed, patient education, by giving the patient the opportunity to choose between her own, and her physician's, preferences, provides her the opportunity to exercise her autonomy. Patient education may help to empower the weak-willed patient who has difficulty forming healthy preferences in the first place by showing her that there are alternatives, and that as an autonomous person, she has the possibility "to change her mind" and how to apply this change. I suggest that this may help to proceed from the precontemplation to the contemplation stage of the Prochaska's model. Of course, this does not mean that the patient will proceed to the next stages (preparation and action) of the model (Reach 2009).

We saw that the difficulty of projection into the future represents an obstacle to patient adherence to long-term therapies. If patient education considers the patient as a person, having her own history, not only past, but also future, it may play a major role in helping the patient to overcome this obstacle by showing her how the therapeutic plan can take into account her projects.

This model shown on Fig. 9.3 may well represent an ideal situation. One should not assume, therefore, that patients must proceed through the cycle, even if they do not want to, or judge it (rationally or irrationally) to be not in their best interest to engage in that cascade. They remain free of deciding for one of the four models. If they are uncertain, medical art may mean being capable of choosing, at the appropriate moment, between various models of the therapeutic relationship. Furthermore, all the time, or sometimes, some persons may wish a paternalistic physician–patient relationship. How should clinicians respond to this patient preference? I suggest that as often as possible, they should adopt an attitude using this ethical pathway: Explaining that there are different alternatives if any, asking for the patient's preference (empathy) and giving their own preference (sympathy) for proposing a given therapy. This again suggests that patient education should at least, at a minima, be present in the process of care. This is why, according to the World Health Organization (1998), patient education may indeed be a necessary part of care: Not only for technical, but also for ethical reasons.

9.10.1 Patient Education and Trust

We suggested that the ethical pathway shown on Fig. 9.3 ends up with a discussion of preferences. As defined by Lewis (1989), a preference is the value a person gives to one over various alternatives and one can better agree on values within a context of trust. An empirical study (Safran et al. 1998) yielded results consistent with the concepts discussed herein: It showed that trust was the variable most strongly associated with patients' satisfaction with their physician, and that the physicians' comprehensive ("whole person") knowledge of patients and patients' trust in their physician were the variables most strongly associated with adherence, patients with higher trust in their physician being significantly more likely to report engaging in eight recommended health behaviors, including exercise, smoking cessation, and safe sexual practices.

Furthermore, a study suggested the following physician behaviors as the factors that determine the trust that a patient has in her HCP. Several qualities of the physician seem to be involved:

> The attribute of technical competency is fairly self-evident, although the physician behavior used by patients to judge technical competency may be quite different from behavior that would be judged by a colleague. Interpersonal competency refers primarily

to communication and relationship-building skills—listening, understanding, providing complete and honest information, and expressing caring. The third domain, agency, is more specific to trust. It refers to acting in the patient's interest—for example, putting the patient's welfare ahead of costs or other considerations. An additional domain, confidentiality, is rarely mentioned by patients and is more weakly associated with the concept of trust as defined by the other domains (Thom 2001).

The emphasis on empathy and communication suggests a link between trust and the ability of the HCP to engage in patient education.

Concerning the involvement of trust in the patient-physician relationship, the overall argument seems to assume that (a) trust can be established between the HCP and the patient, (b) the patient decides that the trust should be established, and (c) is willing to work towards establishing such a trust. Actually, this may not happen: The patient could just as easily decide not to allow the trust to be established, and mandating that trust must be established would be an obvious violation of the patient's autonomy. This points out the importance of the final remark of Shinebourne in his chapter defending a place for paternalism in the patient-physician relationship:

> The doctor must in turn realize that trust (of doctor by patient) has to be earned, gained, must not be abused and should not be assumed – indeed why should it be. (Shinebourne and Bush 1994)

In the model proposed in Fig. 9.3, trust plays its most important role in the last step comparing the patient's and HCP's preferences (i.e., from the interpretative to the deliberative model). Again, the patient remains free to opt for the interpretative model, or to engage or not in the whole cascade, i.e., in patient education. However, the role of trust is also important in the informative model, where the patients has to believe the information given by the physician (providing information is a part of patient education). Indeed, we saw that the formation of a new durable belief (for instance, "I believe that it is good for my health to exercise") is the result of several assessments, including credibility, based on the evaluation of the reliability of the sources at my disposal (Fridja and Mesquita 2000).

This analysis seems therefore to suggest that trust must be somewhat present in a sound patient-physician relationship (even if the physician is obviously in no way entitled to impose it). Actually, the whole medical practice supposes the existence of trust: How can the patient accept to answer the physician questions, the clinical examination of her body, and the proposed therapy? We saw that a relationship of trust cannot be assumed and has to be earned and gained. In addition, the physician, afterwards, has to demonstrate that he/she is trustworthy. Actually, this is a condition of patient's trust in her doctor, according to the concept of "encapsulated interests" developed by Russel Hardin:

> I trust you because I think it is your interest to take my interests in the relevant matter seriously in the following sense. You value the continuation of our relationship and you therefore have your own interest in taking my interests into account. That is, you encapsulate my interests in your own interests. (Hardin 2004)

This importance of trust in the patient-physician relationship also points out the obligation for physicians to reject deception when providing information to the patients. Onora O'Neill in *"Autonomy and Trust in Bioethics"* shows how this obligation has

> ...many implications: It will be expressed in refraining from lying, from false promising, from promise breaking, from misrepresentation, from manipulation, from theft, from fraud, from corruption, from passing off, from impersonation, from perjury, from forgery, from plagiarism, and from many other ways of misleading. More positively, it will be expressed through truthful communication, through care not to mislead, through avoidance of exaggeration, through simplicity and explicitness, through honesty in dealing with others, in a word, through trustworthiness. (O'Neill 2002)

Finally, we also pointed out that a determinant of trust is the clarity of the explanations provided by the HCP. This provides a link between trust and patient education. We can now understand the importance of these words in Maimonides' extraordinary Physician's Prayer: *Let my patients trust me.*

9.11 Conclusion: Mind and Care

It is important to note that patient education's influence is largely based on a cognitive process, which works at the level of our "system 2" of decision making, which is reflective, conscious, controlled, analytic, slow, and cognitively demanding (the Davidson's "having all considered"), rather than on our "system 1", which is unconscious, uncontrolled, heuristic, fast, and cognitively parsimonious (Kahneman 2011; Thaler and Sunstein 2008; Shagai 2013). This distinguishes patient education from a mere manipulation, which would act on system 1.

I suggest that what may be morally acceptable (Shagai 2013) in the context of "nudge"—influencing the behaviors of populations by acting at the level of system 1, which is efficacious precisely because people do not deliberate on their choices, remaining actually free to behave differently (Thaler and Sunstein 2008), would not be acceptable in the case of the care of an individual patient, even if it is aimed at improving the patient's welfare (principle of beneficence), because what would be nothing but a manipulation would violate the fact that the patient is a person, i.e., a being endowed with a reflective mind.

References

Aristotle. Rhetoric. Book 2, Chap. 4; 1381.

Bachimont J, Cogneau J, Letourmy A. Pourquoi les médecins généralistes n'observent-ils pas les recommandations de bonnes pratiques cliniques ? L'exemple du diabète de type 2. Sciences Sociales et Santé. 2006;24:75–102.

Beauchamp T, Childress J. Principles of biomedical ethics. Oxford: Oxford University Press; 2001.

Black DM. Sympathy reconfigured: some reflections on sympathy, empathy and the discovery of values. Int J Psychoanal. 2004;85:579–95.

Cochrane AL. 1931–1971: a critical review, with particular reference to the medical profession. In: Medicines for the year 2000. London: Office of Health Economics; 1979, p. 1–11.

Cogneau J, Lehr-Drylewicz AM, Bachimont J, Letourmy A. Écarts entre le référentiel et la pratique dans le diabète de type 2. Presse Médicale. 2007;36:764–70.

Darwall S. Welfare and rational care, princeton monographs in philosophy. Princeton: Princeton University Press; 2002.

Davidson D. Essays on actions and events. Oxford: Clarendon Press; 2001.

Drass J, Kell S, Osborn M, Bausell B, Corcoran J Jr, Moskowitz A, Fleming B. Diabetes care for medicare beneficiaries. Attitudes and behaviors of primary care physicians. Diab Care. 1998;21:1282–7.

Dworkin G. The theory and practice of autonomy. Cambridge: Cambridge University Press; 1988.

el-Kebbi IM, Ziemer DC, Gallina DL, Dunbar V, Phillips LS. Diabetes in urban African-Americans. XV. In: Identification of barriers to provider adherence to management protocols. Diab Care 1999; 22:1617–20.

Emanuel EJ, Emanuel LL. Four models of the physician-patent relationship. JAMA. 1992;267:221–6.

Engel G. The need for a new medical model: a challenge for biomedicine. Science. 1977;196:129–36.

Frankfurt H. The importance of what we care about. Cambridge: Cambridge University Press; 1988.

Fridja NH, Mesquita B. Beliefs through emotions. In: Fridja NH, Manstead ASR, Bem S, editors. Emotions and beliefs how feelings influence thoughts. Cambridge: Cambridge University Press; 2000.

Grant R, Adams AS, Trinacty CM, Zhang F, Kleinman K, Soumerai SB, Meigs JB, Ross-Degnan D. Relationship between patient medication adherence and subsequent clinical inertia in type 2 diabetes glycemic management. Diab Care. 2007;30:807–12.

Groves JE. Taking care of the hateful patient. N Engl J Med. 1978;298:883–7.

Hardin R. Trust and trustworthiness. New York: Russel Sage Foundation; 2004.

Haynes RB, Taylor DW, Sackett DL. Compliance in health care. Baltimore: Johns Hopkins University Press; 1979.

Hippocrates. Hippocratic writings (trans: Lloyd GER, Chadwick J). Penguin Classics; 1978.

Hojat M. Empathy in patient care: antecedents, development, measurement, and outcomes. New York: Springer; 2007.

Kahneman D. Thinking, fast and slow. Westminster: Allen Lane, Penguin Books; 2011.

Kerr EA, Zikmund-Fisher BJ, Klamerus ML, Subramanian U, Hogan MM, Hofer TP. The role of clinical uncertainty in treatment decisions for diabetic patients with uncontrolled blood pressure. Ann Int Med. 2008;148:717–27.

Leslie CA, Satin Rapaport W, Matheson D, Stone R, Enfield G. Psychological insulin resistance: a missed diagnosis. Diab Spectrum. 1994;7:52–7.

Lewis D. Dispositional theories of values. Proc Aristotelian Soc. 1989;63:113–37.

Miles RW. Cognitive bias and planning error: nullification of evidence-based medicine in the nursing home. J Am Med Dir Ass. 2010;11:194–203.

Miller LV, Goldstein J. More efficient care of diabetic patients in a county-hospital setting. N Engl J Med. 1972;286:1388–91.

O'Neill O. Autonomy and trust in bioethics. Cambridge: Cambridge University Press; 2002.

Parchman ML, Pugh JA, Romero RL, Bowers KW. Competing demands or clinical inertia: the case of elevated glycosylated hemoglobin. Ann Family Med. 2007;5:196–201.

Phillips P. Type 2 diabetes—Failure, blame and guilt in the adoption of insulin therapy. Rev Diabet Stud. 2005;2:35–9.

Phillips LS. It's time to overcome clinical inertia. Ann Int Med. 2008;148:783–5.

Phillips LS, Branch WT, Cook CB, Doyle JP, El-Kebbi IM, Gallina DL, Miller CD, Ziemer DC, Barnes CS. Clinical inertia. Ann Int Med. 2001;135:825–34.

Reach G. Patients nonadherence and healthcare provider inertia are clinical myopia. Diab Metab. 2008;34:382–5.

Reach G. Obstacles to patient education: a transtheoretical analysis. Patient Educ Counsel. 2009;77:192–6.

Reach, G. Obedience and motivation as mechanisms for adherence to medication. A study in obese type 2 diabetic patients. Patient Prefer Adherence. 2011;5:523–31.

Reach G. Patient autonomy in chronic care: solving a paradox. Patient Prefer Adherence. 2013;8:15–24.

Reach, G. Clinical inertia, a critique of medical reason (Foreword by Elster J). Berlin: Springer; 2014a.

Reach, G. Is it possible to improve patient adherence to long-term therapies? Bioethica Forum. 2014b;7:90–7.

Rogers CR. A theory of therapy, personality and interpersonal relationships, as developed in the client-centered framework. In: Koch S, editor. Psychology: a study of science. vol. 3. New York: Mc Graw Hill; 1959.

Safran DG, Taira DA, Rogers WH, Kosinski M, Ware JE, Tarlov AR. Linking primary care performance to outcomes of care. J Fam Pract. 1998;47:213–20.

Schneewind JB. The invention of autonomy. Cambridge: Cambridge University Press; 1998.

Shagai Y. Salvaging the concept of nudge. J Med Ethics. 2013;39:487–93.

Shah BR, Hux JE, Laupacis A, Zinman B, van Walrvaraven C. Clinical inertia in response to inadequate glycemic control. Do specialists differ from primary physicians? Diab Care. 2005;28:600–6.

Shinebourne EA, Bush A. For paternalism in the doctor-patient relationship. In: Gillon R, Lloyd A, editors. Principles of health care ethics. New York: Wiley; 1994.

Tappolet C. Emotions and the intelligibility of akratic actions. In: Stroud S, Tappolet C, editors. Weakness of will and practical irrationality. Oxford: Clarendon Press; 2003.

Thaler RH, Sunstein CR. Nudge: improving decisions about health, wealth, and happiness. New Haven: Yale University Press; 2008.

Thom DH. Stanford trust study physicians, physician behaviors that predict patient trust. J Fam Pract. 2001;50:323–8.

Trope Y, Liberman N. Temporal construal. Psychol Rev. 2003;110:403–21.

Turner BJ, Hollenbeak CS, Weiner M, Ten Have T, Tang SS. Effect of unrelated comorbid conditions on hypertension management. Ann Int Med. 2008;148:578–86.

Wispé L. The distinction between sympathy and empathy: to call forth a concept, a word is needed. J Pers Soc Psychol. 1986;50:314–21.

World Health Organisation. Therapeutic patient education: continuing education programmes for health care providers in the field of prevention of chronic diseases. Copenhagen: WHO; 1998.

Chapter 10
Conclusion: Adherence Generalized

Abstract Not conforming to the principle of foresight, described in this book (Chap. 7), leads to nonadherence. This principle, whose intervention would allow patient adherence, could be called on to help with all incontinent actions which may have potentially serious consequences for our future. This explanation would allow us to define adherence in a general sense, as acceptance to perform certain actions when we believe their consequences could positively influence not only our immediate, but also our long-term, future. Included among these actions is everything that concerns one's health, but also being sober, buckling one's seatbelt, crossing at the crosswalk, avoiding risky situations, and, for some, believing in an afterlife. In sum, the phenomena of adherence and nonadherence concern everything that is vital to us, as opposed to everything that leads us to death. Adherence and nonadherence emerge as the medical counterparts of Eros and Thanatos Freudian principles. Then what answer to give, at the end of this study, to the question: why do we take care of ourselves? We take care of ourselves, if we have a reason to do it, and the reason is the desire—the sustained intention—which confers priority status to the option of arriving healthy and safe at the end of our voyage. We take care of ourselves because it is natural for us to love ourselves just as it is natural to love our children. Here I follow Harry Frankfurt in his celebration of self-love, the highest thing of all, according to Spinoza.

> *Happy the man who, like Ulysses, has traveled well, or like that man who conquered the fleece, and has then returned, full of experience and wisdom, to live among his kinfolk the rest of his life.*
>
> Joachim du Bellay, Regrets

10.1 A Choice Between Two Actions

So, when I refused the cigarette you offered, I first respected the principle of continence that tells me to examine all the contents of the propositional attitudes pertinent to the question being asked. These contents are located in two domains, one where I keep those issues concerning the immediate present, the other containing

© Springer International Publishing Switzerland 2015
G. Reach, *The Mental Mechanisms of Patient Adherence to Long-Term Therapies*,
Philosophy and Medicine 118, DOI 10.1007/978-3-319-12265-6_10

those concerning the future. Next, I put to work my principle of foresight, which tells me to give priority to the arguments found in the domain concerning to the future. This is how a few years ago I was able to come to the realization that it was time to think about my health and once and for all stop smoking. This is a strong expression of the principle of foresight, consisting in making the right resolutions, as part of a habit. Habit, as we have seen, circumvents the need to repeat this exercise day after day, always running the risk, if failing to execute it, of falling off the wagon. *No smoking…*

And if, on the contrary, I accept your cigarette, it means that either my mind lacks the domain of the future (or that it is empty as far as this question is concerned), or that I do not have the principle of foresight, or that the principle is inhibited at the moment, as because of an emotion. The sulky character played by Sabine Azéma in Alain Resnais's *Smoking* takes the packet of cigarettes left on the table after grumbling "what the hell"….

10.2 The Risk of Nonadherence

This example shows that the risk of nonadherence is serious. It is especially so when distant future rewards pale in comparison to immediate rewards. We can understand now what emerged at the beginning of our research, when I enumerated the determinants of patient adherence.

Let us take the example of a typical illness with a risk of nonadherence: A chronic asymptomatic disease requiring a boring preventive treatment, including lifestyle changes such as losing weight, exercising, not smoking. And let us consider first what is being proposed to the patient: The reward of all these tasks is not particularly attractive; it is abstract and often expressed as a negative (to avoid complications), and somewhat tragically, the benefits will never entirely accrue. It is like an insurance policy that one does not completely understand, though the premiums must be paid several times a day and there is no return on the money. As with insurance policies, the chances of the insured-against event happening are far from one hundred percent, and we can gamble on never needing it: Perhaps there will be no fire. (After all, the State requires insurance because this sort of reasoning is so compelling.)

And then there is the incontinent action. It consists of doing nothing, and its rewards (rest, no side effects from the medication, etc.) are immediate, concrete and often visible. The distinction between distant and immediate rewards is certainly not academic: It drives real behavior. We mentioned in the last chapter that Yaacov Trope and Nira Liberman suggested that our mental categorizations tend to give concepts pertaining to the future an abstract, central, important character, expressed for instance by the question: *why*; meanwhile, those concerning the immediate future are given a concrete, more accessorial, less important character, pertaining more to the question: *how* (Trope and Liberman 2003). As we have seen, when a desire is more likely to be satisfied sooner, the motivation to satisfy that desire is typically stronger. In this game, nonadherence has the deck stacked in its favor.

On the other hand, there are circumstances that intensify the motivations pertinent to more distant rewards. We have already mentioned how diabetic women suddenly manifest an extraordinary adherence the moment they decide they want a child, and particularly if they know they are pregnant. One can assume that the diabetic women are no longer acting only for themselves: They sacrifice immediate rewards for their child's later health. But what is more, the object of desire, to bring the pregnancy to term without complications, is concrete and present to the mind. Here, the question is no longer how to avoid complications—a question which is abstract and negative—but how to have a healthy child—which is of course a more concrete and positive matter. Moreover, the payoff, though not immediate, is foreseeable (the woman's due date) which adds to the concreteness of the reward for the required efforts.

Isn't this particular aspect of the favorable influence of pregnancy on adherence explained by its uniqueness, among human actions, the birth of a child? Perhaps rather than seeing pregnancy as a special case, we can view it as a window on a deeper source of strength and motivations. We can quote Hannah Arendt's *Human Condition*:

> The miracle that saves the world, the realm of human affairs, from its normal, "natural" ruin is ultimately the fact of natality, in which the action is ontologically rooted (Arendt 1998, 247).

10.3   Generalization of the Problem

Hannah Arendt gives a superb *general* description of the distinction between two types of rewards, an immediate and a distant one, as it emerges from our study of patient adherence:

> Desire is influenced by what is just at hand, thus easily obtainable – suggestion carried by the very word used for appetite or desire, *orexis*, whose primary meaning, from *orego*, indicates the stretching out of one's hand to reach for something nearby. Only when the fulfillment of a desire lies in the future and has to take the time factor into account is practical reason needed and stimulated by it. In the case of incontinence, it is the force of desire for what is close at hand that leads to incontinence, and here practical reason will intervene out of concern for future consequences. But men do not only desire what is close at hand; they are able to imagine objects of desire to secure which they need to calculate the appropriate means. It is this future imagined object of desire that stimulates practical reason; as far as the resulting motion, the act itself, is concerned, the desired object is the beginning while for the calculating process the same object is the end of the movement (Arendt 1978, 58).

In this book, I have likened patient nonadherence to a *particular* case of *akrasia*. And is it not rather an *exemplary* case of incontinent action, as a manifestation of irrationality, or rather, as I will now show, of a *certain type* of incontinent action?

For philosophers, following Davidson, irrationality manifests itself essentially in three forms: Incontinent actions, self-deception and wishful thinking. Among these types of irrationality, the essence of incontinent actions is precisely the fact that they concern actions, i.e., events that by definition have future consequences, inscribed in temporality.

Among these actions, we can distinguish two types: Those that have somewhat neutral consequences and those that concern one's future. In *Paradoxes of Irrationality*, Davidson illustrates action of the first type with a story taken from Freud. It deals with a man who moves a tree branch lying in a park. Continuing on his way, he thinks that the branch might hurt someone and he is torn between two contradictory desires, to go home or to return to the park and put the branch back in its place. His best judgment tells him to go home. He performs the incontinent action of returning to the park to put the branch back in its place (Davidson 2004, 172–173). Obviously, the consequences of going home or of returning to the park are going to modify his *immediate* future, but it does not have important consequences for his future *in general*. It is an incontinent action of the first type where it is not necessary to engage the principle of foresight.

This principle, whose intervention would allow patient adherence, could be called on to help with all incontinent actions of the second type, which may have potentially serious consequences for the future. This *explanation* would allow us *to define adherence in a general sense*, as acceptance to perform certain actions when we believe their consequences could positively influence not only our immediate, but also our long-term, future. Included among these actions is everything that concerns one's health, but also being sober, buckling one's seatbelt, crossing at the crosswalk, avoiding risky situations, and, for some, believing in an afterlife. We see here again an analogy between health beliefs and religious beliefs: When one Googles the word "observance" (the French word used to speak about adherence to treatment) most references are to observance of religious practices rather than to therapeutic observance. Indeed, the term observance was first created in a religious context.

10.4 Defining Adherence by Its Explanation

In all science there is explanation and understanding; to explain, one must find an understandable link between the different elements one is to explain. (Descombes 2001, 58–66)

We have come back to the idea introduced at the beginning of this inquiry: All these actions—or their opposites—are not only *analogous*, but *phenomenologically homologous* and form, in a sense, a category.

We have often stressed that one must consider separately all the tasks of a medical treatment, each being an action that risks nonadherence, and that nonadherence often is not a general trait, but rather must be examined on a case-by-case basis. However, if clinical experience indeed shows that patients are not usually entirely adherent or nonadherent, it also suggests that phenomena of nonadherence are often linked. Take the example of diabetes: Diabetic patients who smoke usually have the poorest glycemic control (Gulliford and Ukoumunne 2001; Stenstrom and Andersson 2000). Non-smokers accept complex treatments more frequently (Perros et al. 1998). Smokers also come less often to consultations (Dyer et al. 1998). We observed a correlation between adherence to medication and the declaration that one fastens seatbelt in the rear of a car (Reach 2011).

It is while researching how these phenomena are homologous that we have progressively been led to postulate the existence of a principle of foresight, *which explains them generally*. According to George Ainslie, to act according to a principle means to decide at the moment of choice that the action is not isolated, but belongs to a category (Ainslie 1999, 69). We saw that for Robert Nozick, a principle is also a way of seeing an action as symbolically representing a group of actions of the same type, or of a related type, which could be equivalent to the concept of homology, advocated by Jon Elster.

The choice between an additional piece of cake and a diet will be made in favor of the second if it is seen as a choice belonging to a larger category pertaining to adherence in general, which gives priority to everything that guarantees our future and, indeed, protects us from death. The association of the ideas of cake and death may seem exaggerated; it nonetheless conforms to the popular description of gluttony: "You dig your tomb with your teeth."

10.5 Eros and Thanatos

In sum, the phenomena of adherence and nonadherence concern everything that is *vital* to us, as opposed to everything that leads us to destruction and death. In Freudian theory (Freud 2005, 89–92), the death instinct, *thanatos*, is opposed to the principle of life, *eros*. Propelled by the desire to return to Ithaca, where, *happy*, he would find, to wed her again, Penelope, who abolished time by undoing night after night what she had woven during the days of waiting, *Ulysses, has traveled well, and has then returned, full of experience and wisdom, to live among his kinfolk the rest of his life.*

When I try to convince a patient suffering from an asymptomatic chronic illness to take care of herself, how many times am I tempted to say: I would like you to fully enjoy your retirement—live the rest of your life among your kin…

Full of experience and wisdom: The repeated actions, the strength of habit, led him to rationality. And finally, likening the principle of foresight to the Freudian concept of *eros*,—eros, which seeks to force together and hold together the portions of living substance (Freud 1990, 75), we may suppose that next to mental states (conscious, or as I have said, unconscious) a place may be specifically reserved for the unconscious to explain nonadherence: The place taken up by the death instinct in us. Perhaps this is why pregnancy has the unique virtue of encouraging adherence.

To refuse to be adherent is to surrender to that pure wickedness towards oneself evoked by Dostoyevsky's character *in Notes from the Underground*:

> a bad character, and so completely unreasonable, since we have long known that one follows the other.

And this is done doubtlessly both knowingly while being conscious of one's irrationality and unconsciously by giving rein to one's death instinct.

The death instinct is described by Freud as

> the inborn human inclination to '*badness*', to aggressiveness and destructiveness, and so to cruelty as well (italics are mines) (Freud 2005, 114).

10.5.1 Why Do We Take Care of Ourselves? The Two Meanings of Why

The 'why' of the question 'why do we take care of ourselves' can be understood in two ways. First interpretation: why in the sense of *for what reason or cause*? To phrase the question in this way is to justify the resort to the Causal Theory of Action, but we saw that it stops at the principle of continence. Proposing an additional principle involving a time dimension as necessary to explain how one may be adherent means considering a second interpretation of the word why: *why* do we take care of ourselves—that is, *with what goal*?

In other words, the causal theory which explains an action by its reason lets us understand how it is possible to be adherent or nonadherent, and answers the question *why* in the first sense. The principle of foresight, which from then on makes apparent that the problem of adherence exists within a temporality, answers the question *for what*. We saw that the same context led Ainslie to propose a different explanation of 'weakness of the will': The change of preferences linked to the passage of time.

In fact, the two meanings of *why* appear in the evolution of Davidson's conception of the Causal Theory of Action. Paul Ricœur, analyzing *Actions and Events*, shows how Davidson himself includes time in his analysis of action fifteen years after initially developing his theory:

> It did not escape Davidson's attention that intending-to presents new and original features, precisely the orientation toward the future, the delay in accomplishing, even the absence of accomplishing, and at least silently, the implication of an agent. [...] But as soon as we consider actions, which, as we say, take time, anticipation operates during the entire unfolding of the action. Is there any sort of extended gesture that I could accomplish without anticipating in some sense its continuation, its completion or interruption? Davidson himself considers the case in which, writing a word, I anticipate the action of writing the next letter while still writing the present letter. How could we fail to recall, in this connection, the famous example or reciting a poem described in Augustine's *Confessions*? The entire dialectic of *intentio* and *distentio*, constitutive of temporality itself, is summed up here: I intend the poem in its entirety while reciting it verse by verse, syllable by syllable, the anticipated future transiting through the present in the direction of a completed past (Ricœur 1995, 81–82).

Then what answer to give, at the end of this study, to the question: why do we take care of ourselves? We take care of ourselves, if we have a reason to do it, and the reason is the desire—the sustained intention—which confers priority status to the option of arriving healthy and safe at the end of our voyage. *Happy is he who like Ulysses...*
According to Ricœur:

> judgment that pleads solely in favor of an action is one thing; judgment that engages action and is sufficient for it is something else again, The formation of an intention is just this unconditional judgment (Ricœur 1995, 82).

Here we see again the notion of an unconditional judgment which leads to action, as opposed to a *prima facie* judgment, a notion which was used by Davidson as his first explanation of *akrasia*.

10.5.2 Foresight, Prudence, and Happiness

The evocation of Ulysses' happiness leads us to clarify the relationship between the principle of foresight and the virtue of *prudence*, Ulysses' quality *par excellence*. The Latin word meaning to foresee, *providere*, has the same root as prudence, the *phronesis* of the Greeks, 'the greatest good' according to Epicurus, which makes us renounce certain pleasures:

> While therefore all pleasure because it is naturally akin to us is good, not all pleasure should be chosen, just as all pain is an evil and yet not all pain is to be shunned. It is, however, by measuring one against another, and by looking at the conveniences and inconveniences, that all these matters must be judged (Epicurus, *Letter to Menoeceus*).

and Aristotle discusses it at length in *Nicomachaen Ethics* before turning to *akrasia*:

> Well, it is thought to be the mark of a prudent man to be able to deliberate rightly about what is good and advantageous for himself, not in particular respects, e.g. what is good for health or physical strength, but what is conductive to the good life generally (Aristotle, *Nicomachean Ethics*, VI, 5, 1).

Aristotle's notion of happiness, *eudemonia*, is understood as a long-term proposition, and in fact can only be truly appreciated at the end of life:

> For one swallow does not make spring, nor does one fine day; and similarly one day or a brief period of happiness does not make a man supremely blessed and happy (Aristotle, *Nicomachean Ethics*, I, 7, 16).

This quote seems to indicate that prudent people are those who do not count on Providence, but generally give priority to the future in the context of a vital project where health is an essential element for today, but only one element of happiness.

I take care of myself because I want to arrive safe and healthy at the end of my voyage. So the impatient ones who say: "But when I am old, I will no longer be myself, why should I deprive myself today," are perhaps rational in their choice: As Derek Parfit put it (Parfit 1984), my concern for my future may correspond to the degree of connectedness between me now and myself in the future. Since connectedness is nearly always weaker over longer periods, I can rationally care less about my further future.

However, Parfit noted elsewhere that we should consider our future selves, who will perhaps no longer be us, as we regard our children or friends. In this case, protecting one's future would be similar to protecting our children or friends, and not grabbing at a moment of pleasure. According to Parfit, such pleasure-seeking at the expense of children, friends, or one's own future self may be rational, but is morally wrong, even if one does in fact bear the consequences herself (quoted in Frederick 2003).

I take care of myself because it is natural for me to *love myself* just as it is natural to love my children. Harry Frankfurt, in a fundamental text that we will soon revisit, notes that self-love should not have the negative aspect that we are sometimes tempted to attribute to it:

> After all, are we not told [...] that we should love our neighbors *as we love ourselves*? That injunction does not sound like a warning against self-love. It neither declares nor implies that we should love others *instead* of loving ourselves. Indeed, it does not in any way suggest that self-love is an enemy of virtue, or that it is somehow discreditable to hold the self dear. On the contrary, the divine command to love others as we love ourselves might even be taken to convey a positive recommendation of self-love as an especially helpful paradigm – a model or ideal, by which we ought seriously to guide ourselves in the conduct of our practical lives (Frankfurt 2004, 77).

The dying king's desperate cry "I love myself!" at the end of Ionesco's play loses its comic aspect and gains a tragic dimension: Here we hear the king's ardent desire to persevere in his desire to take care of himself, to love himself, to persevere in his being, as Spinoza would say: He sees that faced with death, which was announced to him at the end of the play, all his desires have become an illusion.

But as long as death is only an abstraction, while I can still forget that my life is also a play that will end one day, and because I can continue to love myself, I take care of myself: I take care of myself, of my existence, projecting myself into the future.

Thus, this care of self is much more than care for the body and goes beyond a strictly medical activity. Michel Foucault analyses Socrates' injunction 'take care of yourself' to Alcibiades:

> Can we say that the doctor takes care of himself when, because he is ill, he applies to himself his knowledge of the art of medicine and his ability to make diagnoses, offer medication, and cure illnesses? The answer is, of course, no. What is it in fact he takes care of when he examines himself, diagnoses himself, and sets himself a regimen? He does not take care of himself in the meaning we have just given to 'himself' as soul, as soul-subject. He takes care of his body, that is to say of the very thing he uses. It is to his body that he attends, not to himself. The first distinction then is that the tekhne of the doctor who applies his knowledge to himself and the tekhne that enables the individual to take care of himself, that is to say take care of his soul as subject, must differ as to their ends, objects, and natures (Foucault 2005, 57–58).

To take care of oneself: To care for one's soul as subject or, if one prefers, as the *person* that we are.

In another passage of the *Hermeneutics of the subject*, Foucault shows

the intertwining of the practice of the self with the general form of the art of living (tekhne tou biou), an integration such that care of the self was no longer a sort of preliminary condition for an art of living that would come later. The practice of the self was no longer that sort of turning point between the education of the pedagogues and adult life, and this obviously entails a number of consequences for the practice of the self. First, it has a more distinctly critical rather than training function: it involves correcting rather than teaching. Hence its kinship with medicine is much more marked, which to some extent frees the practice of the self from pedagogy. Finally, there is a privileged relationship between the practice of the self and old age, and so between the practice of the self and life itself, since the practice of the self is at one with or merges with life itself. The objective of the practice of the self therefore is preparation for old age, which appears as a privileged moment of existence and, in truth, as the ideal pint of the subject's fulfillment. You have to be old to be a subject (Foucault 2005, 125–126).

Then let us be clear: In the words of Nicolas Postel-Vinay and Pierre Corvol,

the quest for perfect health doubtlessly has pitfalls and the first would be to expect of medicine a happiness that it is incapable of giving (Postel-Vinay and Corvol 2000, 265).

But it is not here a question of the myth of perfect health, but simply of taking care of oneself to preserve one's health, which is one of the elements of happiness.

The relationship between health and happiness appears clearly, as notes Jean-François Mattéi (2001, 60), in the origin of the Greek word for health, *ugieia*, which

comes from two Sanskrit roots: *su, 'good', in Greek *eu*, like in *eudaimonia*, 'happiness', and *giyʷ-es-*, 'life', which Greek takes up in the words *bios* and *zen*. *Ugieia* is the 'good' or 'happy' life, 'life in good health', one could even say 'in full light', as the radical *giyʷ-es-*, which is found in *zen*, 'to live' and also in Zeus, 'the living' or 'luminous', with the genitive *Dios* giving the Latin *Deus*, 'god', and also *dies*, 'day'. According to the Greeks, life is given to the one who comes to the light before going to linger in the shadowy kingdom of Hades, like Achilles in the Odyssey. Health is thus a gift of the gods, because it is life itself, in the plenitude of light, the creative energy as incarnated, even before the god of medicine Asklepios, by the goddess of health, Hygieia, known to us through Pausanias.

According to this etymology, health is light, this light that the sick person is desperately seeking for, described by Marcel Proust in the first lines of the *Search of Lost Time*

Nearly midnight. The hour when an invalid, who has been obliged to start on a journey and to sleep in a strange hotel, awakens in a moment of illness and sees with glad relief a streak of daylight shewing under his bedroom door. Oh, joy of joys! it is morning. The servants will be about in a minute: he can ring, and some one will come to look after him. The thought of being made comfortable gives him strength to endure his pain. He is certain he heard footsteps: they come nearer, and then die away. The ray of light beneath his door is extinguished. It is midnight; some one has turned out the gas; the last servant has gone to bed, and he must lie all night in agony with no one to bring him any help.

10.5.3 Eros

> I am not being treated and never have been, though I respect medicine and doctors. What's more, I am also superstitious in the extreme; well at least enough to respect medicine. [...] No, sir, I refuse to be treated out of wickedness. Now you will certainly not be so good as to understand this.
>
> Dostoyevsky, Notes from the Underground

> Self-approval is in reality the highest object for which we can hope
>
> Spinoza, *Ethics*, Book 4, Proposition 52

> ...and the whole of humanity, in space and in time, is one immense army galloping beside and before and behind each of us in an overwhelming charge able to beat down every resistance and clear the most formidable obstacles, perhaps even death.
>
> Bergson, Creative Evolution

If the obstacle of death can be overcome, life will be victorious. We may indeed liken the principle of foresight, which makes this miracle possible, to the Freudian *eros*.

In order to understand how we can accept taking care of ourselves, we have postulated the existence of a unifying principle, which overcomes the division of one's will, a division which is its weakness. Owing to this principle—the principle of foresight—we mitigate the opposing arguments. And by applying this principle, we reestablish a unity of the will. It is the principle of foresight, which, in the choice leading us to take care of ourselves (or not), allows us to complete a potentially endless discussion. We are then capable of wholeheartedly adhering to the desire to care for ourselves.

Wholehearted adherence to one's desires thanks to a unified will defines the profound significance of self-love, described by Harry Frankfurt in a text which is worth quoting in full[1]: This may be Spinoza's self-approval.

> If ambivalence is a disease of the mind, the health of the mind requires a unified will. That is, the mind is healthy – at least with respect to its volitional faculty – insofar as it is wholehearted. Being wholehearted means having a will that is undivided. The wholehearted person is fully settled as to what he wants, and what he cares about. With regard to any conflict of dispositions or inclinations within himself, he has no doubts or reservations as to where he stands. He lends himself to his caring and loving unequivocally and without reserve. Thus his identification with the volitional configuration that define his final ends is neither inhibited nor qualified.
>
> This wholehearted identification means that there is no ambivalence in his attitude toward himself. There is no part of him – that is, no part with which he identifies – that resists his loving what he loves. There is no equivocation in his devotion to his beloved. Since he cares wholeheartedly about the things that are important to him, he can properly be said to be wholehearted in caring about himself. Insofar as he is wholehearted in loving those things, in other words, he wholeheartedly loves himself. His wholehearted self-love consists in, or is exact constituted by, the wholeheartedness of his unified will.

---

[1] Frankfurt (2004). © 2004 Princeton University Press. Reprinted by permission of Princeton University Press.

To be wholehearted *is* to love oneself. [...]

One thing in favor of an undivided will is that divided wills are inherently self-defeating. Division of the will is a counterpart in the realm of conduct to self-contradiction in the realm of thought. A self-contradictory belief requires us, simultaneously, both to accept and to deny the same judgment. Thus it guarantees cognitive failure. Analogously, conflict within the will precludes behavioral effectiveness, by moving us to act in contrary directions at the same time. Deficiency in wholeheartedness is a kind of irrationality, then, which infects our practical lives and renders them incoherent.

By the same token, enjoying the inner harmony of an undivided will is tantamount to possessing a fundamental kind of freedom. Insofar as a person loves himself – in other words, to the extent that he is volitionally wholehearted – he does not resist any movements of his own will. He is not at odds with himself; he does not oppose, or seek to impede, the expression in practical reasoning and in conduct of whatever love his self-love entails. He is free in loving what he loves, at least in the sense that his loving is not obstructed or interfered with by himself.

Self-love has going for it, then, its role in constituting both the structure of volitional rationality and the mode of freedom that this structure of the will ensures. [...] Perhaps Spinoza is right. Loving oneself may well be the 'highest' or the most important thing of all (Frankfurt 2004, 95–98).

To take care of ourselves is to love ourselves, and to not do so is pure wickedness. Now we can understand the cry of the tortured hero of *Notes from the Underground.*

Mind and Care
Take care of yourself
Care for yourself
Love yourself

Vale: Take care.

References

Ainslie G. The dangers of willpower. In: Elster J, Skog O-J, editors. Getting hooked, rationality and addiction. Cambridge: Cambridge University Press; 1999.

Arendt H. The life of the mind. Harvest Books; 1978.

Arendt H. Human condition. Chicago: The University of Chicago Press; 1998.

Aristotle. Nicomachean ethics.

Davidson D. Paradoxes of irrationality. In: Problems of rationality. Oxford: Clarendon Press; 2004.

de Spinoza B. Ethics.

Descombes V. The mind's provisions (trans: Schwartz SA). Princeton: Princeton University Press; 2001.

Dostoyevsky F. Notes from the underground (trans: Pevear R, Volokhonsky L). Vintage; 1994.

Dyer PH, Lloyd CE, Lancshire RJ, Bain SC, Barnett AH. Factors associated with clinic non-attendance in adults with type 1 diabetes mellitus. Diabet Med. 1998;15:339–43.

Epicurus. Letter to Menoeceus.

Foucault M. Hermeneutics of the subject: lectures at the Collège de France 1981–82 (trans: Davidson A). Basingstoke: Palgrave Macmillan; 2005.

Frankfurt HG. The reasons of love. Princeton: Princeton University Press; 2004.

Frederick S. Time preference and personal identity. In: Loewenstein G, Read D, Baumeister RF, editors. Time and decision. New York: Russel Sage Foundation; 2003.

Freud S. Beyond the pleasure principle. Norton Library; 1990.

Freud S. Civilization and its discontents. Norton Library; 2005.

Gulliford MC, Ukoumunne OC. Determinants of glycated hemoglobin in the general population: associations with diet, alcohol and cigarette smoking. Eur J Clin Nutr. 2001;55:615–23.

Mattéi JF. Platon et le modèle rationnel de la santé. In: L'Utopie de la santé parfaite, Colloque de Cerisy, Sfez L, editor. P. U. F., La politique éclatée; 2001.

Parfit D. Reasons and persons. Oxford: Clarendon Press; 1984. p. 313–4.

Perros P, Deary IJ, Frier BM. Factors influencing preference of insulin regimen in people with type 1 (insulin-dependent) diabetes. Diab Res Clin Pract. 1998;39:23–9.

Postel-Vinay N, Corvol P. Le retour du Dr Knock. Éditions Odile Jacob; 2000.

Proust M. In search of lost time (trans: Scott Moncrieff CK).

Reach G. Obedience and motivation as mechanisms for adherence to medication. A study in obese type 2 diabetic patients. Patient Prefer Adherence. 2011;5:523–31.

Ricœur P. Oneself as another (trans: Blamey K). Chicago: University of Chicago Press; 1995.

Stenstrom U, Andersson P. Smoking, blood glucose control, and locus of control beliefs in people with type 1 diabetes mellitus. Diab Res Clin Pract. 2000;50:103–7.

Trope Y, Liberman N. Temporal construal. Psychol Rev. 2003;110:403–21.

By the Same Author

Clinical Inertia, A Critique of Medical Reason, Forewords by Jon Elster and Joël Ménard, Springer, 2014.

© Springer International Publishing Switzerland 2015   199
G. Reach, *The Mental Mechanisms of Patient Adherence to Long-Term
Therapies*, Philosophy and Medicine 118, DOI 10.1007/978-3-319-12265-6

Index

A

Abelhauser, 101
Abstinence, 7, 75, 121
A condition of satisfaction, 80
Action, 1, 9–11, 22, 24, 26, 28–32, 35, 39, 40, 44–52, 55–59, 61, 62, 64–75, 77, 79–84, 86, 89, 91–99, 101–104, 117–119, 121, 122, 128, 129, 134, 142, 146, 147, 150, 153, 159, 169, 170, 188–193
Addiction, 7, 25, 100, 118, 145, 152
Addison, 131
Adherers, 4, 18
Agency, 29, 55, 58, 86, 182
Agent, 18, 22, 31, 32, 35, 36, 43, 46, 47, 49, 51, 52, 63–65, 68, 69, 77, 84, 92–99, 101–105, 111, 113, 116, 117, 119, 133, 134, 141, 142, 146, 158, 159, 169, 192
Age of foresight, 107, 131
AIDS, 3, 4, 74, 145
Ainslie, 30, 79, 109, 113, 115, 118, 132, 133, 191, 192
Ajzen, 23
Akrasia, 12, 30, 89, 91–93, 95–102, 104, 107, 118, 119, 131, 134, 142, 169, 170, 189, 193
Akratic, 92, 96, 104, 119, 170
Alcohol, 1, 21, 29, 75, 85, 89, 112, 114, 117, 121, 128, 133, 154
Alcoholics anonymous, 117, 158
All things considered, 10, 51, 69, 89, 92, 95–97, 102, 104, 105, 119, 120, 129, 169, 170
Amygdala, 126
Analogy, 8
Anderson, 17, 157
Anger, 38, 49, 60–62

Anouilh, 41
Anscombe, 46, 47
Antabuse, 75, 114
Anxiety, 38, 44, 60–62, 77, 78, 174
Anxious, 18, 72, 84
Appointment, 1, 2, 6, 19, 71, 74, 140
Apter, 27, 122, 125
Arendt, 189
Aristotle, 2, 10, 50, 70, 91, 95, 100, 130, 174, 193
Assal, 100
Asthma, 4, 6, 155
Augustine, 132, 192
Augustus, 154
Authority of the first person, 140
Autonomous, 6, 97, 102, 139, 150–155, 157, 161, 162, 164, 180
Autonomy, 92, 149–158, 161–164, 171, 178–180, 182
Aymé, 139

B

Background, 41, 57, 59, 76
Balint, 145, 158
Baudelaire, 98
Baumeister, 45, 85, 117
Beauchamp, 151, 178
Becker, 22
Behavior, 1–3, 5, 8–11, 15, 18, 19, 22–31, 41, 44–46, 48, 52, 56, 59, 61, 64, 65, 67, 70, 71, 74, 75, 79, 80, 85, 89–92, 96, 98–100, 103, 108, 111, 112, 114, 116, 117, 120, 121, 125, 129, 130, 134, 150, 161, 168, 173, 178, 179, 181, 188
Behaviorist, 28, 29, 41, 75, 133

© Springer International Publishing Switzerland 2015
G. Reach, *The Mental Mechanisms of Patient Adherence to Long-Term Therapies*, Philosophy and Medicine 118, DOI 10.1007/978-3-319-12265-6

Belief, 11, 16, 24, 27, 29–31, 35–43, 40–51, 55, 56, 58, 59, 62, 67, 68, 78–80, 83, 84, 93, 98, 101–104, 119, 120, 122–124, 128, 129, 139–148, 151, 153, 154, 162, 163, 168, 170, 182, 190, 197
Beneficence, 150, 161, 162, 174, 179, 183
Benefits, 9, 22, 26, 46, 70, 73, 83, 100, 133, 167, 188
Bentham, 123
Bergson, 136
Bickel, 111, 117, 121, 126
Black, 173, 174
Bloch-Lainé, 145, 146
Boettiger, 128
Boredom, 61, 119
Bratman, 80, 81

C
Caligula, 101
Callias, 159, 160
Camus, 101
Canguilhem, 123, 130, 131, 135, 149, 159
Care, 1, 3–7, 9, 11, 12, 15, 16, 19, 28, 55–57, 59, 60, 68, 72, 74, 75, 77–82, 99, 107, 118–120, 124, 131, 134, 135, 139, 155, 157, 159, 160, 162–164, 167–169, 172–175, 178–181, 183, 187, 191–197
Carnap, 95, 102
Catechol-O-methyltransferase, 128
Causal, 15, 27, 31, 40, 44, 47–49, 51, 56, 64, 97, 120, 142, 192
Cavell, 97
Chapman, 108, 111, 118
Childress, 151, 178
Choice, 5, 10, 11, 17, 18, 21, 24, 30, 42, 44, 68, 75, 77, 78, 91–94, 104, 107, 108, 111–113, 115, 118, 120, 121, 125–127, 133, 134, 150, 151, 156, 161, 180, 187, 191, 193, 196
Christakou, 126
Chronic, 20, 107
Chronic diseases, 3, 10, 12, 20, 153, 155, 160, 164
Cigarette, 1, 5, 7, 10, 18, 21, 27, 65, 70, 78, 99, 113–115, 117, 118, 121, 152, 187, 188
Clinical inertia, 12, 148, 149, 164, 167–176, 178
Cognition, 38, 78, 128
Cognitive, 7, 11, 17, 22, 25, 36, 38, 39, 44, 45, 59, 82, 101, 102, 114, 124–126, 128, 132, 154, 183, 197
Compliance, 1, 2, 74, 132, 169

Compulsive, 90, 91, 98, 99, 121
Content, 29, 31, 35–42, 56, 67, 68, 80, 81, 103, 104, 120, 129, 130, 134, 141, 142, 144, 146, 160, 161
Coping, 24, 73, 101
Correia, 104
Corvol, 195

D
Damasio, 31, 75, 112, 125, 131
Darwall, 173–177, 179, 180
Davidson, 30, 41, 43, 47–52, 57, 58, 60, 68, 70, 77, 79, 81, 89, 92–98, 101, 102, 104, 118–120, 129–132, 140–142, 169, 190, 192, 193
Dawking, 135
Dawkins, 135
Death, 22, 63, 101, 118, 133, 136, 159, 191, 192, 194, 196
Decision, 18, 22, 26, 38, 44, 63, 67, 71, 75, 77, 79, 80, 84, 85, 91, 93, 94, 107, 108, 113, 114, 116–119, 122, 128, 130, 132, 133, 150, 151, 153, 155–157, 161, 178, 183
Deliberative model, 156
Delpla, 141
Democritus, 73, 100
Denial, 11, 26, 61, 62, 90, 100–104, 108, 129, 130
Dennett, 135
Depreciation, 108
Descartes, 31, 32, 112, 125
Descombes, 27, 32, 41, 140–142, 159, 160, 163, 190
Desire, 11, 21, 29, 35–38, 40, 41, 44, 47–51, 55, 56, 58–61, 65, 67, 68, 70, 72, 77, 79, 83, 84, 90, 93, 94, 97, 98, 100, 102–104, 107, 108, 110, 112–116, 119, 120, 122–124, 129, 134, 135, 139–142, 144, 145, 148, 149, 151–155, 157, 158, 160–163, 187, 189–191, 193, 194, 196
De Souza, 43
Device, 113, 130
DiClemente, 25, 79
Diabetes, 2–4, 6, 7, 16, 20, 56, 63, 68, 69, 74, 99, 123, 144, 147, 157, 159, 169, 170, 191
Diet, 1, 7, 10, 23, 25, 26, 29, 38, 42, 47, 57, 63, 67, 69, 74, 85, 90, 100, 112, 113, 116–118, 133, 191
Direction of fit, 37, 104
Discounting, 108–114, 122, 148, 154
Dispositional-functionalist, 39

Doctor, 1, 2, 5, 6, 9, 11, 12, 16–19, 24, 26, 38, 55, 57, 61, 71, 72, 77, 89, 90, 92, 116, 139, 140, 143–145, 147, 148, 150, 155, 156, 158, 159, 161, 162, 167–179, 182, 194
Dokic, 40
Dostoevsky, 192
Dostoyevsky, 105, 196
Dreyfus, 76
Drinking, 25, 26, 70, 154
Drugs, 1, 20, 112
Du Bellay, 187
Dupuy, 42
Durkheim, 73, 123
Dworkin, 152, 153, 163

E

Elster, 8, 18, 30, 31, 38, 39, 44, 45, 60, 110, 112–114, 116–118, 124, 128, 130, 158, 191
Emanuel, 155, 176
Emotional, 6, 25, 39, 59–61, 104, 116, 125, 126, 128, 132, 134, 145, 148, 149, 161, 172, 174, 177
Emotions, 11, 18, 20, 29, 31, 36–38, 43–46, 55, 58–63, 73, 75, 77, 78, 85, 99, 103, 104, 116, 117, 119, 126, 128, 129, 139, 142, 148, 149, 153, 154, 157, 163, 167, 168, 170, 172, 174, 175, 178
Empathy, 145, 172, 173, 176
Empowerment, 157, 160
Engel G, 162, 176
Engel P, 10, 29, 30, 39, 40, 42, 43, 80, 123, 141, 142, 146, 148
Epicurus, 193
Epilepsy, 3
Eros, 191, 196
Essays on actions and events, 94
Evidence-based medicine, 169
Exercise, 1, 2, 7, 21, 25, 26, 29, 47–50, 64, 67, 69, 74, 84, 93, 94, 100, 115, 118, 123, 133, 136, 153–155, 178, 180–182, 188
Exponential, 108, 109, 111, 115

F

Fear, 18, 19, 31, 36, 37, 43, 44, 47, 60–63, 68, 77, 90, 102, 119, 122, 129, 170–174, 179
Feeling, 38, 39, 45, 56, 65, 84, 122, 146, 148, 149, 172–174, 178
Feldman, 148
Financial, 4

Fishbein, 23
Foresight, 10, 12, 107, 119–121, 124, 129–134, 136, 143, 147, 158, 188, 190–193, 196
Forgetfulness, 6
Foucault, 160, 194, 195
Frankfurt, 134, 152, 153, 163, 175, 194, 196, 197
Frederick, 109, 194
Freud, 97, 132, 190–192
Freudian, 97, 98, 103, 132, 142, 191, 196
Fridja, 42, 182
Future, 9, 12, 25–27, 47, 51, 56, 60, 63, 72, 77, 79, 81, 101, 107–110, 112–116, 118–125, 127–131, 133, 135, 143, 146–148, 154, 158, 163, 164, 167, 168, 170, 173–176, 178, 181, 187–194

G

Gfeller, 100
Gide, 124, 130
Gjelsvik, 117, 118, 120, 121
Gout, 3, 20
Guidelines, 167, 168, 171, 174
Guilt, 60, 62, 122

H

Habit, 7, 12, 24, 29, 41, 70–77, 79, 82, 83, 85, 86, 90, 99, 118, 134, 188, 191
Habitual actions, 73
Happiness, 39, 121, 193, 195
Hardin, 182
Hate, 38, 146
Haynes, 2, 5, 179
Health, 4–6, 9, 15, 18–20, 22, 24, 27, 29, 30, 46, 55, 56, 59–61, 63, 65–67, 70, 72, 79, 81, 83, 84, 86, 89, 100, 108, 111, 115–117, 119, 121–124, 135, 145, 147–151, 154–160, 162, 168, 169, 171, 172, 176, 180–182, 187–190, 193, 195, 196
Health belief model, 9, 15, 22, 23, 80, 108
Health literacy, 6, 7
Heil, 40
Hempel, 95, 102
Herman, 118, 121
Herrnstein, 109
Heuristics, 17, 92
Hippocampus, 126
Hippocrates, 147, 175
Hojat, 172
Holism, 41, 42, 45, 131, 140

Holistic, 24, 40, 51, 76, 130, 140, 141, 144, 148
Holton, 83, 84, 133
Homer, 114
Homology, 7, 8, 191
Hope, 9, 36, 37, 60, 96, 99, 143, 144, 170,
    180, 196
Horwitz, 4, 133
Hospital, 19
Hume, 40, 73
Hyperbolic, 109–111, 113
Hypercholesterolemia, 3
Hypertension, 3, 4, 20, 169
Hypothyroidism, 3, 20

I
Ignorance, 6
Illness, 1, 2, 16, 18–20, 22, 24, 29, 32, 60–63,
    73, 74, 77, 81, 90, 100, 101, 103, 107,
    108, 119, 123, 129, 136, 145, 146, 155,
    158, 162, 163, 170, 188, 191, 195
Impatient, 108, 110, 114, 125, 128, 154, 193
Inaction, 169
Incontinence, 10, 20, 89, 91–93, 96, 119, 189
Incontinent, 10, 89, 91–95, 97–100, 102, 103,
    107, 117, 119, 122, 129, 130, 169,
    187–190
Informative model, 156, 176, 178, 182
Ingelfinger, 155, 161
Intention, 8, 10, 11, 23–25, 36, 46, 47, 50, 57,
    67, 71, 79–84, 96, 101, 108, 132, 146,
    171, 177, 187, 193
Intentional, 6, 28, 29, 35–39, 41, 46, 55, 57,
    59, 60, 67, 75, 80, 82, 83, 91, 96, 101,
    150
Intentionalist, 11, 29, 55, 58, 139, 151, 153
Intention in action, 81–83
Intentionality, 11, 12, 28, 35, 38, 47, 67, 81, 83
Intermediate rewards, 116
Interpretation, 1, 4, 11, 24, 81, 90, 131, 135,
    141, 143, 160, 192
Interpretative model, 156, 157, 162, 176, 180,
    182
Intertemporal choice, 11, 30, 118
Intertemporal choice, 11, 30
Ionesco, 194
Irrational, 6, 9, 43, 45, 46, 80, 90, 96, 98, 99,
    102–104, 114, 118, 120, 121, 132
Irrationality, 89, 96–105, 119, 189, 190, 192, 197

J
Johnson, 5, 111, 117
Jones, 122

Joseph, 77, 126, 128

K
Kacelnik, 125, 130
Kahneman, 17, 92, 183
Kalenscher, 127
Kantian, 163
Katz, 161
Kendall, 131
Knock, 63
Knowledge, 2, 7, 8, 11, 15, 16, 18, 29, 35, 37,
    40, 42, 48, 55, 56, 58, 90, 95, 129, 130,
    139, 140, 146, 147, 153, 157, 159, 163,
    181, 194

L
Laplantine, 17, 123
Laws, 31, 35, 50–52, 124, 130, 141, 150
LeDoux, 31, 126
Leibniz, 35
Leriche, 123
Leventhal, 22, 24, 25, 46, 59, 65, 74, 125
Lewis, 152, 181
Liberman, 175, 188
Life, 3, 9–11, 18, 20, 29, 30, 40, 41, 48,
    60–63, 69, 74, 75, 107, 118, 122–124,
    132, 135, 136, 140, 149, 151, 160, 170,
    177, 187, 191, 193–196
Linguistic, 16
Livet, 43, 44, 60–62, 73, 77, 78, 104, 116, 119,
    148, 149, 154
Locke, 64, 83
Loewenstein, 30, 44, 99
Logical, 11, 30, 31, 36, 40, 50, 97, 104, 131, 168
Loi Huriet, 151
Losonsky, 42
Love, 38, 47, 60, 141, 145, 146, 174, 187,
    194, 197

M
Mammography, 25, 121
Manuck, 126, 127
Marcel Proust, 195
Maslin, 36, 40
Mattéi, 195
Mechanisms, 11, 28, 29, 31, 39, 40, 45, 46,
    48, 51, 75, 76, 89, 103, 114, 124, 125,
    128–130, 164
Medication, 1–6, 9, 11, 17–21, 24, 29, 58, 59,
    71, 72, 74, 75, 81, 82, 92, 103, 118,
    121, 122, 127, 133, 180, 188, 191, 194

Medication event monitoring systems, 57
Medicine, 16, 17, 19, 20, 46, 105, 129, 131, 133,
    135, 147, 149, 151, 158, 159, 161–163,
    167, 168, 176, 178, 179, 194–196
Meichenbaum, 2, 15, 17, 19, 21
Mele, 28, 30, 67–70, 72, 74, 101, 102
Memes, 135
Mental, 4, 11, 15, 27–29, 31, 35–41, 43,
    45, 47–49, 51, 52, 55, 58, 59, 61, 67,
    79, 80, 83, 85, 89, 97, 98, 108, 120,
    122–124, 129, 131, 132, 139–141, 145,
    151, 153, 154, 157, 163, 188, 191
Mental events, 35, 36, 52
Mental states, 11, 15, 27–29, 31, 36–41, 43,
    45, 48, 49, 51, 55, 58, 59, 67, 79, 80,
    97, 108, 120, 122, 123, 139, 140, 151,
    153, 154, 157, 163, 191
Mesquita, 42
Miles, 169
Miro, 70
Mischel, 124–126, 132
Morality, 94, 114
Moro, 143
Mortality, 3, 4, 20, 63, 101
Motivational, 12, 38, 39, 55, 59, 63, 64, 67,
    68, 72, 111, 146
Motivational force, 12, 67, 68, 72, 111, 146
MRI, 126, 128
Myopia, 12, 112, 164, 175

N
Neuroscience, 31
New year's resolutions, 115
Nonsmoking, 117
Nordenfelt, 123
Norms, 23, 96, 135, 141, 149
Nozick, 42, 45, 68, 112–115, 124, 128–130,
    134, 135, 191
Nudge, 183
Nuttin, 77

O
O'Connell, 100
Ogien, 50, 51, 94, 95, 98, 99, 132
O'Neill, 183
Organ transplantation, 4
Osteoporosis, 3
Osler, 161

P
Pacherie, 81

Pain, 3, 22, 24, 35, 37–39, 55, 59, 73, 131,
    147, 193, 195
Paradoxical, 89, 90, 93, 102, 159, 178, 179
Parfit, 109, 193, 194
Partitioning of the mind, 96
Partitioning of the mind, 97, 102, 103, 119,
    120, 131, 142
Paternalistic model, 18, 155, 157, 176, 180
Patient, 1–7, 9–12, 15–21, 23–26, 28–32, 44,
    46, 47, 52, 55–63, 65, 67–69, 71–83,
    89–93, 95, 99–101, 103, 105, 107, 108,
    110, 116–121, 123, 124, 127, 128,
    130–132, 134, 139–151, 154–164,
    167–183, 187–191
Patient and agent, 31
Patient education, 7, 143, 158, 160, 163
Pears, 10, 44, 102, 103
Penelope, 113, 191
Personal rules, 115, 133
Pharo, 39, 65, 124
Phillips LS, 167, 168, 170
Phillips P, 170
Philosophers, 10, 11, 28, 30, 36, 48, 64, 89,
    91, 92, 169, 173, 190
Philosophical, 6, 10, 11, 29, 36, 46–48, 79, 84,
    90, 93, 131, 134, 151, 152, 160, 180
Philosophy, 10, 11, 29, 30, 36, 40, 48, 52, 89,
    123, 140, 141
Philosophy of mind, 11, 36, 89, 140
Physiology of mind, 31, 36
Pathophysiological, 130, 131, 134
Pierret, 74
Pigeons, 110, 124, 127
Pima Indians, 63, 128
Pity, 38
Plato, 91
Pleasure, 5, 27, 39, 55, 59, 65, 73, 75, 93, 99,
    104, 109, 116, 117, 123, 132, 145, 149,
    151, 172, 193, 194
Polivy show, 118
Popper, 42
Postel-Vinay, 195
Practical syllogism, 50, 51, 94, 95
Precommitment, 113, 114, 116
Prefrontal cortex, 126
Pregnancy, 3, 21, 189, 191
Prescription, 1, 3, 5, 6, 11, 18, 20, 71, 92, 143,
    158, 167, 169
Pride, 37, 45, 60, 64
Primary reason of action, 68
Primitive actions, 49, 68
Principle of charity, 141, 142
Principle of continence, 89, 95–98, 102, 107,
    119–121, 129–132, 134, 187, 192

Principle of foresight, 121, 132, 136, 143, 188, 196
Prior intention, 81–83
Prochaska, 25, 26, 75, 78–80, 83, 85, 108, 145, 181
Proposition, 35, 36, 38, 40, 51, 102, 110, 142, 193
Propositional attitudes, 35, 36, 39–42, 56, 96, 131, 141, 142, 148, 187
Prospect theory, 17
Proust, 64, 65, 76, 79, 83, 84, 134
Prudence, 193
Psychological, 10, 11, 22, 25, 27, 28, 30, 31, 46, 62, 73, 90, 91, 100, 108, 132, 141, 145, 151, 152, 155, 158, 160, 163, 170, 171, 173
Psychological models, 11, 27, 30, 46
Psychologist, 10
Psychology, 9, 10, 21, 22, 29–31, 36, 64, 69, 140, 151
Public health, 136
Puzzle, 35, 42, 43, 45, 46, 61, 104, 129, 144

R
Rachlin, 70
Ramsey, 11, 30, 40, 129, 147
Rational, 9, 23, 43, 59, 60, 78, 93, 95, 97, 98, 103, 104, 113, 114, 116, 121, 129, 132, 133, 135, 141, 193, 194
Rationality, 7, 42–44, 97, 102–104, 120, 121, 128–130, 132, 135, 141, 191, 197
Rats, 75, 110, 126
Ravaisson, 72
Reach, 4, 7, 16–18, 20, 73, 77, 82, 99, 118, 120, 127, 143, 146, 175, 178–181, 191
Reactance, 18
Recklessness, 90, 91
Reflective, 99, 151, 152, 159, 160, 183
Regret, 36, 37, 44, 45, 60, 98, 99, 113
Relapses, 26
Resolution, 12, 62, 67, 71, 79, 84, 90, 108, 117
Resources, 19, 23, 55, 59, 90, 151
Restaurant, 68, 112, 117
Reversal theory, 15, 22, 27
Revision, 43, 44, 61, 62, 77, 78, 104, 119, 149, 154
Reward, 10, 110, 111, 113, 114, 116, 117, 120, 124–126, 133, 154, 188, 189
Ricœur, 163, 192, 193
Risk, 3, 9, 11, 17, 19, 22, 40, 44, 74, 76, 79, 90, 99, 113, 115, 121, 122, 124, 125, 133, 141, 143, 147, 153, 154, 159, 163, 177, 188

Rogers, 145, 172
Romains, 63
Rorty, 30
Ruse, 113, 114
Russell, 36, 42

S
Sadness, 38, 101
Safe sex, 15, 23
Saint Augustine, 132, 192
Saint Martin, 148
Saint Paul, 98
Sartre, 28, 80
Schizophrenia, 4
Searle, 37, 38, 41, 44, 57, 59, 76, 79–82
Seatbelt, 4, 18, 118, 180
Secondary order mental states, 59
Selective serotonin reuptake inhibitors, 127
Self-control, 12, 69, 70, 72, 74, 100, 113, 117
Self-efficacy, 22, 26
Self-love, 12, 194, 196, 197
Self-regulatory model, 15, 22, 24
Seneca, 79
Shagai, 183
Shame, 37, 38, 45, 60, 62
Shinebourne, 162, 182
Sirens, 69, 113
Skills, 11, 29, 41, 55, 58, 74–76, 157, 178, 182
Skinner, 70, 75, 129
Smith, 90
Smoke, 1, 8, 104, 114, 133, 136, 149, 152, 191
Smokers, 7, 18, 63, 78, 89, 99, 111, 133, 191
Smoking, 4, 7, 18, 22, 25, 27, 29–31, 39, 44, 63, 67–70, 75, 79, 85, 90, 100, 104, 114–117, 121, 122, 133, 149, 151, 152, 181, 188
Socrates, 50, 91, 139, 194
Somatic markers, 75, 131
Spinoza, 110, 145, 194, 196, 197
Stroud, 92
Surprise, 37, 60, 61, 72, 73
Syllogisms, 50, 94
Symbolism, 134
Sympathy, 27, 168, 172–181

T
Tappolet, 92, 104, 170, 175
Tauber, 162, 163
Temporal, 10, 12, 81, 101, 120, 121

Temptation, 21, 26, 67, 83, 85, 98,
    113–115
Thalamus, 126
Thaler, 109, 183
Thanatos, 191
Theories of reasoned action and of planed
    behavior, 15, 22
Theory of interpersonal behavior, 15, 22, 24
Theory of planned behavior, 23
Therapeutic agency, 52, 55, 56, 108
Therapeutic alliance, 12, 143, 145, 158, 162
Thom, 182
Threat, 22, 24
Time, 3, 6, 10, 16, 17, 19, 20, 23, 25,
    26, 29, 45, 48, 51, 56, 57, 61,
    62, 67, 69–73, 75, 76, 78, 80, 83–85,
    90, 92, 96, 98–102, 107–113, 115,
    118–122, 124–126, 131, 133, 136,
    139, 143, 147–152, 154–159, 161,
    163, 164, 168, 174, 179, 181, 188,
    189, 191, 192, 197
Top-down model of adherence, 132
Transtheoretical model of change, 15, 22, 25,
    26, 108
Triandis, 24
Tristan Bernard, 124
Trope, 175, 188
Trust, 144, 146, 147, 162, 164, 178,
    181–183
Turk, 2, 15, 17, 19, 21
Tversky, 17, 92
Typology, 7

U
Ulysses, 69, 74, 113, 114, 130, 158, 162, 187,
    191, 193
Uncertainty, 17, 18, 101, 122, 146, 147, 155,
    161, 169

V
Valery, 139
Visceral, 11, 38, 39, 42, 44, 59
Volition, 47, 64, 71, 76, 83, 84
Volitional, 64, 65, 196, 197
Vulnerability, 16, 22, 26

W
Wallace, 121
Watson, 30, 90, 91, 98, 100
Weak, 79, 91, 118, 127, 134, 146, 157, 180
Weakness of the will possible, 93–95
Weakness of will, 12, 30, 79, 84, 85, 89, 91,
    92, 94, 100, 118
Weight, 25, 29, 42, 47–50, 55, 57, 59, 61, 63,
    64, 67, 69, 76, 84, 89, 92, 93, 98, 99,
    114, 115, 118, 134, 151, 153, 170, 188
WHO report, 5
Wholehearted, 196, 197
Wickedness, 105, 192, 196, 197
Willpower, 64, 67, 83–86, 90, 91, 100, 134
Wispé, 172, 173
Wittgenstein, 40, 48
World health organization, 4, 149, 181

Printed in the United States
By Bookmasters